LUFTWAFFE COLORS

Volume 1 1935-40

LUFTWAFFE COLORS

Volume 1 1935-40

K. A. Merrick
With line illustrations by the Author

Color paintings by Geoffrey Pentland

Arco Publishing Company Inc
New York

PUBLISHED BY ARCO PUBLISHING COMPANY INC,
219 PARK AVENUE SOUTH, NEW YORK,
NEW YORK 10003

Library of Congress Catalog Card Number 74-14272
ISBN 0-668-03652-4

FIRST PUBLISHED IN 1973

THIS EDITION PUBLISHED IN 1975 BY ARRANGEMENT WITH KOOKABURRA TECHNICAL PUBLICATIONS PTY LTD, 81 POTTER ST, DANDENONG 3175, VICTORIA, AUSTRALIA
PRINTED IN HONG KONG

INTRODUCTION

It is not the aim of the author to describe the campaigns of the *Luftwaffe*, the deeds of its personnel, nor the development of its equipment—other and more able pens have already done this. The intention however is simply, and that term is used relatively, to provide a chronological detailed history of the development of camouflage and markings as applied to aircraft of the *Luftwaffe* from its prenatal days to its demise.

Many works, both large and small, already exist concerning this subject; some cover certain aspects in detail while others are broader in scope but offer less depth. It is sincerely hoped that the current series of books will provide a detailed analysis of this fascinating and often frustrating subject. This work arose from a strong personal desire to have available a ready reference that could be consulted quickly about a particular point, any point, and which would supply a satisfying, specific and comprehensive answer.

A great deal of documentation, both official and personal, has been carefully and critically examined. This has been well supported, in turn, by the memories of many who served in the *Luftwaffe*. If it is to be truly comprehensive, any serious historical survey must examine all previously published material relating to the subject. One such source is deserving of special mention; the comprehensive series of books produced by Karl Ries. The seemingly endless variety and scope of the photographic material contained within them provides a rich field for investigation. That I have chosen to disagree on some points with Herr Ries's conclusions, and indeed the conclusions of some others, is done with what I hope approaches complete objectivity, and is motivated purely by my interpretation of the surviving evidence. In such cases I hope that my own viewpoint will, if nothing else, provide food for further thought and investigation.

Every care has been taken to verify the facts assembled within these covers. I offer my sincere thanks for the patience of so many who have been consulted, with frustrating monotony, on a multitude of points to establish that the conclusions reached are indeed facts and not just opinions.

Doubtless there will be criticisms levelled at some of the photographs selected for this first volume. I would ask only that the reader realise that the very early period under examination was marked by the almost continuous metamorphosis of color schemes and markings, both subtle and obvious. To record as many of the known variations as possible, it has been necessary in some cases to utilise photographic material as poor in technical quality as it is rare in detail. Hopefully the serious student of this fascinating subject will accept these limitations in the interests of establishing an accurate historical record.

Any such work as this is heavily dependent upon photographs, and I am endebted to Herr H. Regel of the Bundesarchiv for his patience in fulfilling my frequent requests for specific material. A similar debt of thanks is extended to H. Birett and R. Zwillsperger of the Deutsches Museum, also to M. Dalancourt of the Etablissement Cinématographic Des Armées, E. Hine of the Imperial War Museum, the Smithsonian Institute, Dr. Buschhorn of Vereinigte Flugtechnische Werke-Fokker GmbH, J. Pöllitsch of Junkers Flugzeug- und Motorenwerke AG, Messerschmitt-Bölkow-Blohm AG, J. Lassere of Icare, Kosmos, Ing H-J. Klein of His Aero Col Par, R. Harrison of Candid Aero Files, W. Hesz of IPMS Austria, H. Obert, Lt-Col A. Krüger, S. Haggard, F. Smith, E. Tusch, N-A. Nilsson, B. Nicolas, H. Jonas, H. de Weerd, P. Malone, B. Robertson, I. Primmer, Z. Titz, M. Pegg, H. Nowarra, N. Parnell, and Real Photographs Co Ltd.

Information has been provided by so many that to try and record the names of all concerned would be impractical. The following institutions and personnel are, however, owed a significant debt of thanks: Col Rougevin-Baville of the Musée de l' Air, the Militärgeschichtliches Forschungsamt der Bundeswehr, the Air Historical Branch of the Ministry of Defence (UK), Herbol-Werke, the Royal Netherlands Air Force Museum, the Swedish Military Archives, Heinkel Flugzeugbau GmbH, H. Falk, W. Wagner, T. Römer, J. Rosenstock, H. Volker, C. Schaedel, W. Berge, J. Kitchens, G. Duval, G. Turner, D. Carpenter, W. Witts, H. Procter, d'E. Darby, P. Anderson, T. Seay and E. McDowell.

Several people are deserving of a special note of gratitude for their particular contribution to this work. Herr F. Hydrich, a veteran pilot who first saw action with the *Legion Condor*, greatly assisted in the clarification and identification of aircraft colors from the fascinating and long-neglected Spanish theatre of operations; Tom Hitchcock, himself an ardent student of this subject, generously and most unselfishly made available to me his own extensive researches into *Luftwaffe* camouflage; Hartmut Tusch uncomplainingly put up with a steady flow of German documents which he so painstakingly translated for me; David Vincent tirelessly tracked down many of the ex-*Luftwaffe* personnel concerned in the preparation of this work and subsequently interviewed them in his usual methodical manner. Finally Geoff Pentland is sincerely thanked for his patience in translating my detailed black-and-white illustrations into a series of meticulous color illustrations to his usual high standard.

To all who have written anything before on this subject I give my thanks and, like them, I hope that this work will inspire others to pursue the subject to even greater depths.

Highbury, South Australia
January 1973

Kenneth Merrick

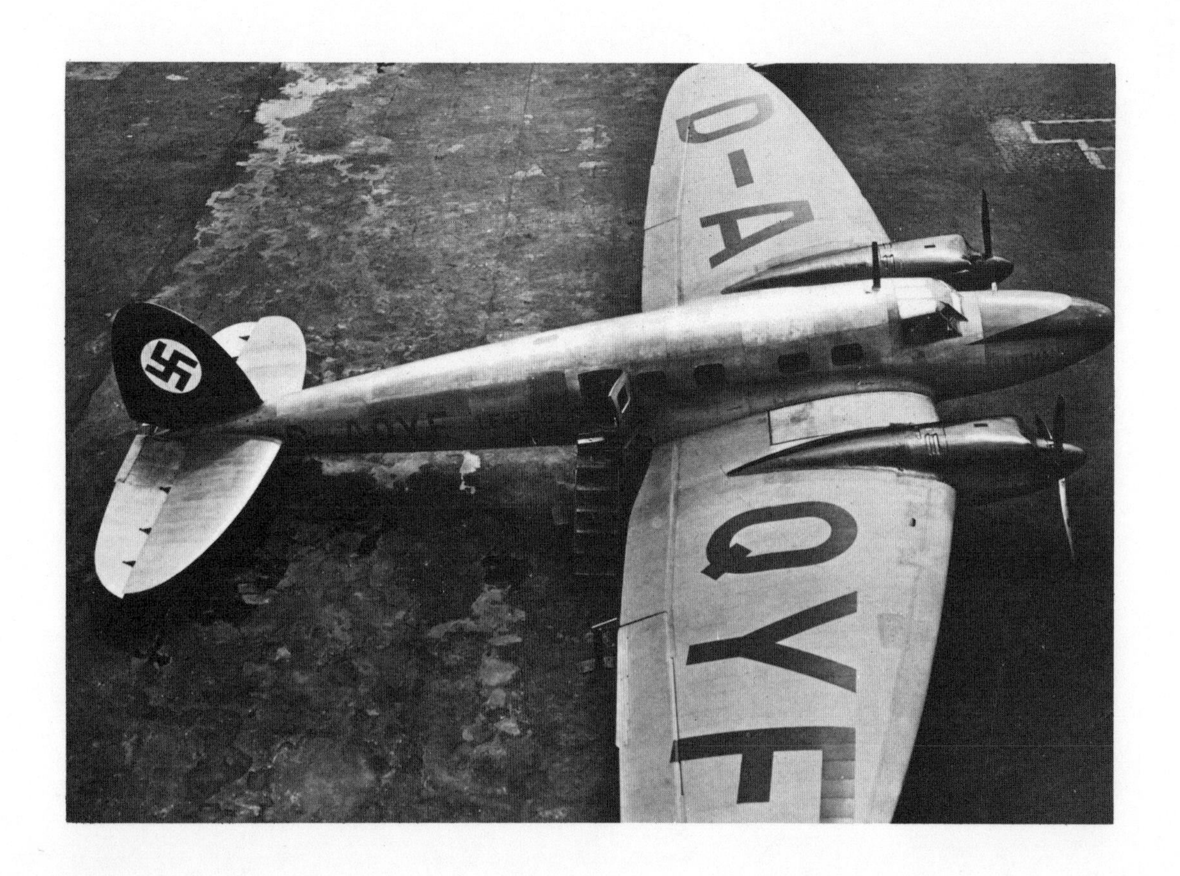

Luftwaffe in embryo...

CONTENTS

Its own career paralleling that of the *Luftwaffe* from its secret beginnings until its ultimate demise, the versatile Ju 52 served in a multitude of rôles and on every known battlefront. It is seen above in its original rôle as a pure transport aircraft. Finished in overall silver colored aluminium paint with its characteristic neat black trim, it bore the civil registration D-2702 in black.

CHAPTER 1
REBIRTH OF AN AIR FORCE

The Treaty of Versailles marked the official termination of the Imperial German Air Force. With its demise history closed its pages, at least temporarily, on its colorfully-bedecked aircraft, and a generation's heraldry met the ignominious fate inevitably meted out to a vanquished foe. That same treaty was however, ambiguous enough to allow Germany to retain a small aircraft industry, and a group of military officers took that opportunity to lay the foundations of a new air force—a *Luftwaffe*.

Experimentation with purely military types of aircraft was well under way in 1933 and by the following year several new machines had made their appearance. The comprehensive range of duties that these types were capable of fulfilling left little doubt about the far-sighted planning involved.

At this time the Heinkel He 45 and He 46 were in production for long- and short-range tactical reconnaissance while the pure fighter rôle was filled by another Heinkel product, the He 51. For bombing, the Dornier Do 11C had been ordered into production after its appearance in late 1933, but powerplant delays seriously hampered its service introduction and the Junkers Ju 52/3m ge rapidly outpaced it in numbers. The Ju 52 had already proven itself as a civil airliner and its adaptation for the auxiliary bomber rôle was as equally straightforward. A more definitive generation of bombers was represented by the Dornier Do 17, albeit still a civil prototype in 1934, the Junkers Ju 87 undergoing construction, and the Heinkel He 111 which was destined to take to the air in the spring of 1935. The training rôle was filled by new types such as the Arado Ar 66 and Focke-Wulf Fw 44 which were built to supplement the existing range of older machines.

Following the close of hostilities in 1918, a simple civil registration system was introduced using the prefix letter D, for *Deutschland,* followed by a numerical sequence commencing from one. With the coming to

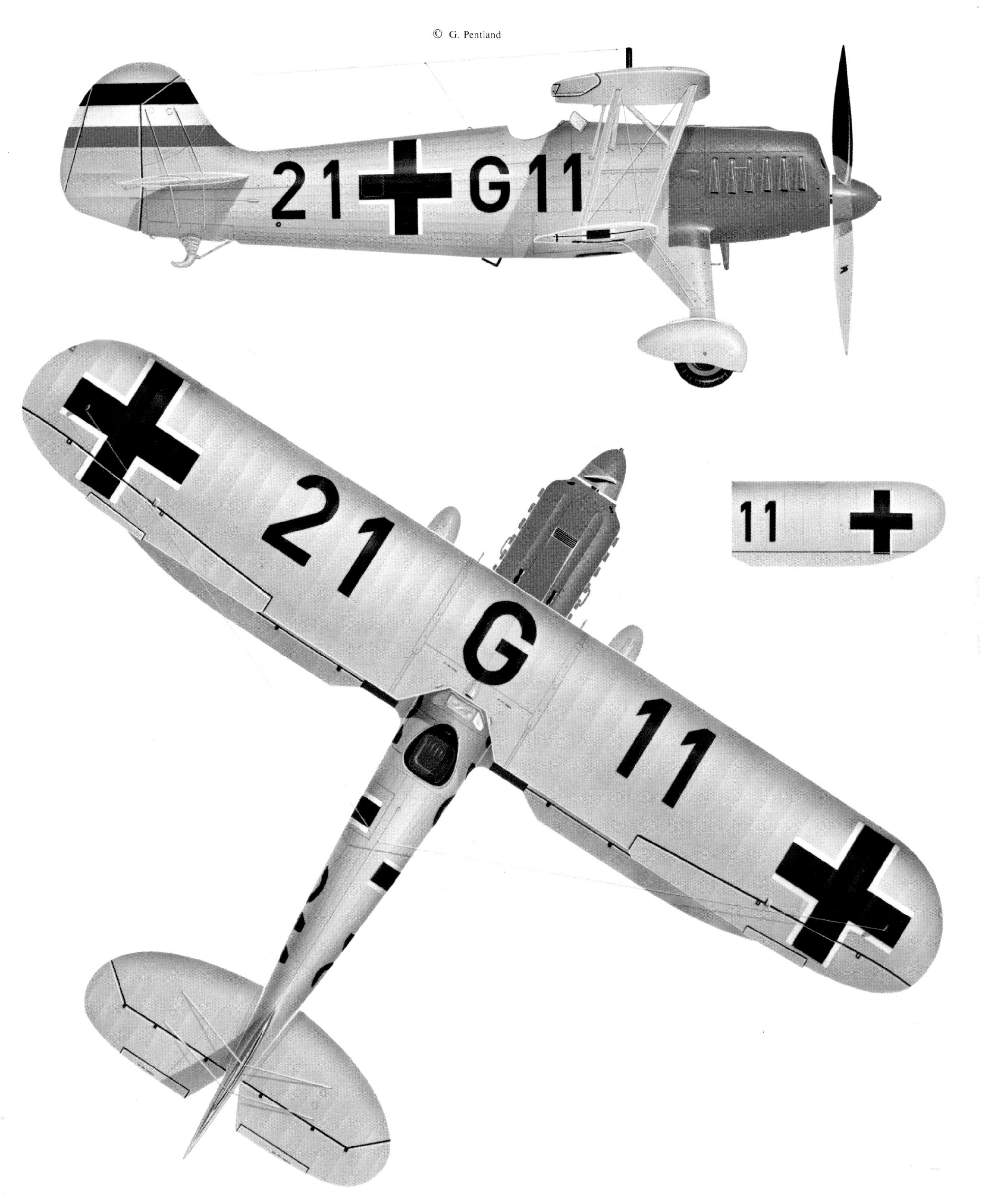

A He 51 of I /JG 132, finished in overall standard pale greenish-grey, at the time when military markings and codes were first introduced for the new *Luftwaffe*. This is the first known major variation from the newly issued standard for *Luftwaffe* aircraft. Note that while codes and markings were all proportionally correct, they were grossly oversize. This resulted in the fuselage *Balkenkreuze* being no longer diametrically opposite each other. The spacing of the codes and markings on both upper and lower wing surfaces was also abnormal, only the individual letter G being correctly positioned on both the upper centre section and on the centre line of the bottom of the fuselage, between the wing roots. The port side of the fin and rudder carried the stylised National Socialist blood banner. The red nose section was the unit's traditional color.

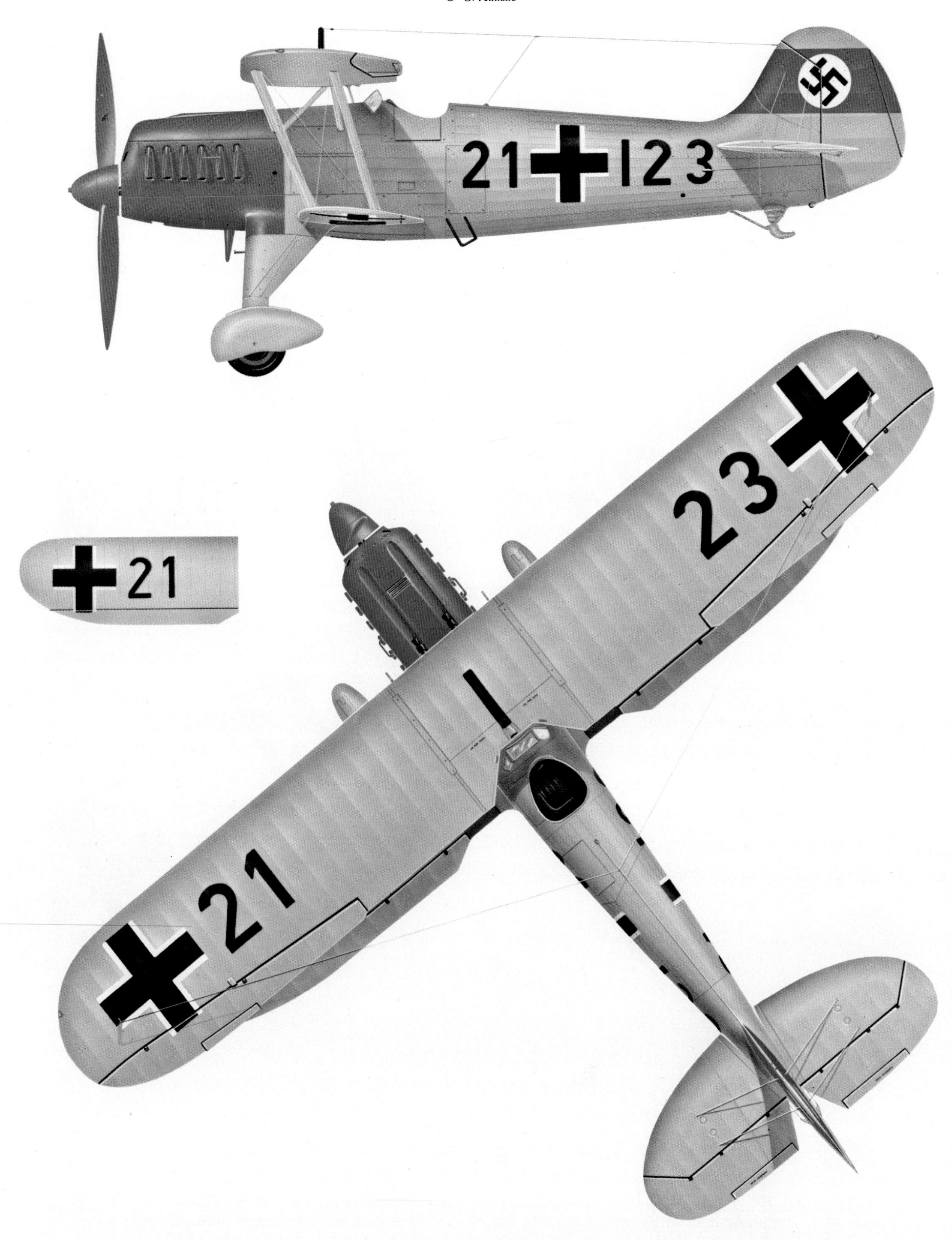

An interesting comparison is possible between this He 51 of II /JG 132 and the one previously illustrated. This aircraft carried not only markings and codes of correct proportions, size and spacing, but also bore the *Hakenkreuz* emblem on both sides of the fin and rudder. Note that the *Balkenkreuze* on both aircraft had no thin black edging to the white surrounds, a feature of early military markings as explained in the main text.

Illustrating the revised style of civil registration introduced in Spring 1934, this He 70 carried a four-letter group starting with the letter U which identified the class C, single-engined registration group.

The He 111C-04 *Königsberg* also bore a class C registration but in this case the first letter, A, denoted that it was a multi-engined aircraft. It was common practice to name aircraft after German cities, famous people, etc, hence the name on the fuselage. In similar civil guise the He 111C-03 *Köln* and the He 111 V4 *Dresden* carried out clandestine photographic reconnaissance duties in 1937 under the control of the *Kommando Rowehl.*

The He 111C-02 *Leipzig* seen here with a group of tri-motored Ju 52s in the background. Note that the registration letters overlapped the control surfaces on the He 111 due to the low aspect ratio of the wing. (See Diagram 1 for details of the proportions.) This photograph was taken in 1936 after the deletion of the earlier-style starboard side fin and rudder red, white and black striping.

A final example from the landplane class C registration group was the four-engined Bv 142 V1, *Kastor*. All the types illustrated, or their direct derivatives, served in due course with the *Luftwaffe*.

A Dornier *Wal,* forerunner of the Do15 *Militär-Wal* 33. It bore a class C seaplane registration.

power of the National Socialist Party in 1933, this system was replaced by a new one which retained the prefix letter D but used a four-letter group in place of numerals. The old system was phased out over a period of some two to three years.

The new registration system divided aircraft up into various specific categories based on the number of people carried, the flying weight and, in a special class, the number of engines and the flying weight. The system also differentiated between landplanes and seaplanes. The first letter of the four-letter registration group identified the actual class to which the aircraft belonged, as follows:

LANDPLANES

Class	Registration Group	Personnel and Weight
A_1	D-Y . . .	1 person, all-up weight 500 Kg
A_2	D-E . . .	1 to 3 persons, all-up weight 1,000 Kg
B_1	D-I . . .*	1 to 4 persons, all-up weight between 1,000 and 2,500 Kg
B_2	D-O . . .	1 to 8 persons, all-up weight between 2,500 and 5,000 Kg
C	D-U . . .	Single-engined } all-up weight in excess of 5,000 Kg
	D-A . . .	Multi-engined } all-up weight in excess of 5,000 Kg

SEAPLANES

Class	Registration Group	Personnel and Weight
A_1	D-Y . . .	1 person, all-up weight 600 Kg
A_2	D-E . . .	1 to 3 persons, all-up weight 2,200 Kg
B	D-I . . .	1 to 4 persons, all-up weight 5,000 Kg
C	D-A . . .	Multi-engined, all-up weight in excess of 5,500 Kg

* This registration group was originally listed as D-J . . . but was apparently never used, as all photographic examples of aircraft in this class are registered in the D-I . . . range.

The remaining three letters in the group were issued in strict alphabetic sequence from AAA to ZZZ. Thus the following blocks of registration letters were available:

D-YAAA	to	D-YZZZ
D-EAAA	to	D-EZZZ
D-IAAA	to	D-IZZZ
D-OAAA	to	D-OZZZ
D-UAAA	to	D-UZZZ
D-AAAA	to	D-AZZZ

The group D-IAAA to D-IZZZ was reserved for the so-called experimental aircraft, ie types intended principally for military purposes.

The first true production fighter for the clandestine *Luftwaffe* was the Arado Ar 65 which had appeared, in prototype form, in 1931. Delivery of the definitive model, the Ar 65E, began in late 1933. These aircraft equipped the Berlin-Staaken-based *Staffeln* of the *Reklamefliegerabteilung* (Publicity Flying Department).

Consistent with the new regulations, the Ar 65Es were given registrations from the D-IAAA range, applied to both sides of the fuselage, the upper surface of the top wings and the lower surface of the bottom wings. The spacing and size of these markings were in accordance with the specific geometric proportions shown in Diagram 1. The lettering was painted in black to give maximum contrast with the overall silver finish of the aircraft.

In addition to this new registration system the National Socialist Government had also introduced a fin and rudder marking consisting initially of two insignia. On the port side was a stylised representation of the National Socialist Party flag, a black *Hakenkreuz* within a white disc bordered by a red square. The entire marking was restricted to the rudder. The starboard side however, bore a horizontal band across the rudder, this being divided horizontally into the three national colors of black, white and red.

The port side insigne was employed in the restricted form for only a few months before being extended across both fin and rudder. The starboard side marking was then extended to match. The *Hakenkreuz* was positioned centrally, the depth of the red background being determined by the available height between the top of the tailplane and the top of the rudder. This was

After the official revelation of the *Luftwaffe* in March 1935 its aircraft retained their civil registrations for some months. These two photographs were reputedly taken in August 1935 and show a mixed line-up of pre-production He 51A-0s and He 51A-1s. The overall finish was the pale greenish-grey then being introduced to replace the old silver finish. The starboard side fin and rudder striping was still in use at this time.

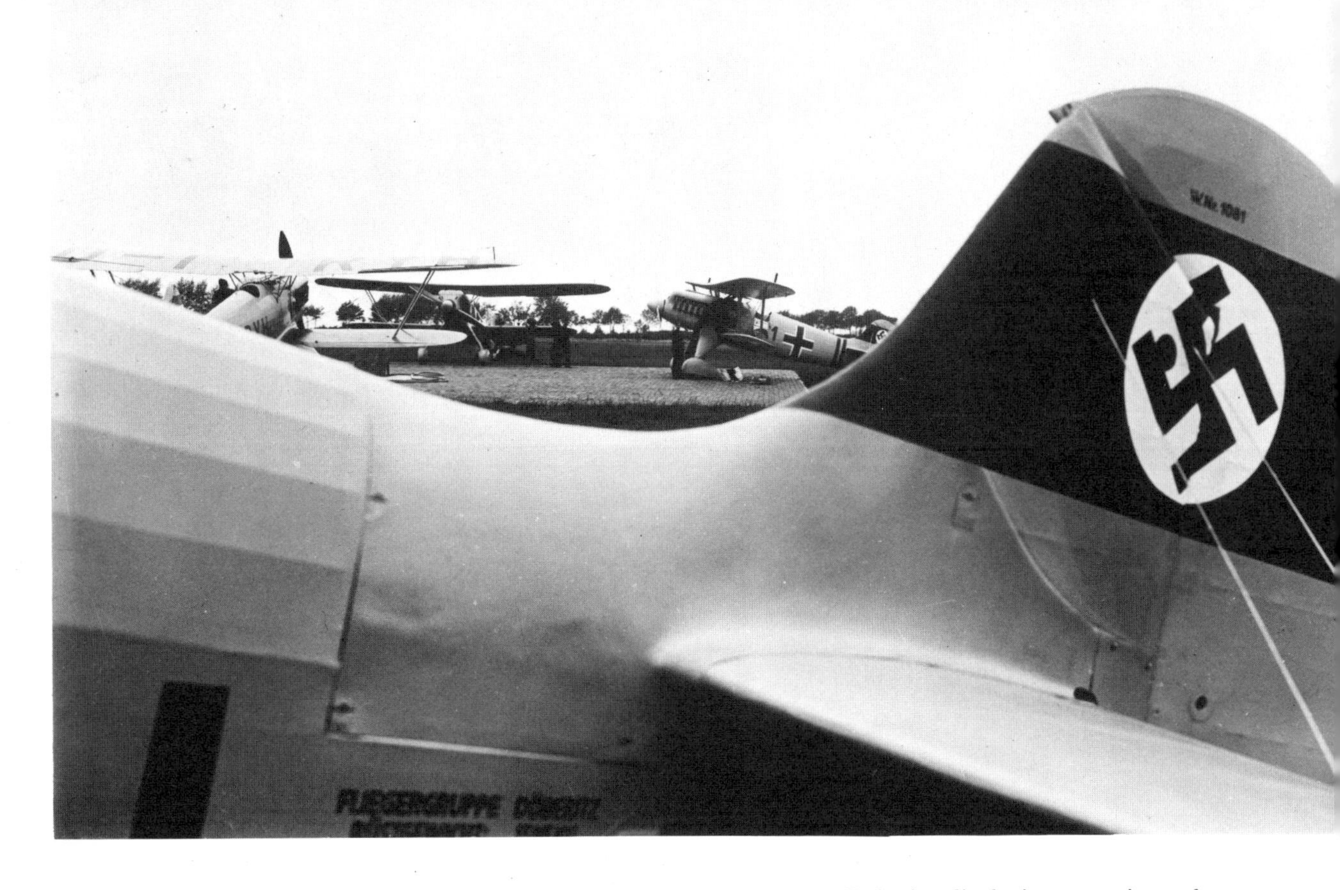

The transition in progress. Above. A group of He 51s of I/JG 132 at Döberitz displaying a variety of markings. On the left is a He 51A-1 still wearing its class B1 registration D-IRYH; in the centre is a He 51 with a very dark propeller hub and blades instead of the usual pale grey and lightly stained wood common to this type. On the right is another He 51 but this time wearing the new military markings. The *Balkenkreuze* on the bottom wing lower surfaces covers the full chord. This last aircraft also wears a type of code marking not previously listed. In the foreground is the rear section of another He 51 with a civil registration and portion of the weights and loading legend is just visible. What appears to be a black disc on the *Hakenkreuz* is actually an inspection port. Below. A closer view of the unusual markings on W Nr 1084. The code A1 was possibly a unit-originated form of code marking to identify a *Gruppe* or *Staffel* leader's aircraft. The two narrow black bands around the rear fuselage are thought to have been a form of air-to-air marking. Although not commonly seen, these, or similar types of markings, have been noted on other aircraft up to 1939/40. This photograph also provides a good example of the early form of *Balkenkreuz* with no black edging to the white surround.

done in accordance with the prescribed set of proportions shown in Diagram 2.

In the spring of 1934 a pre-production batch of nine Heinkel He 51A-0 fighters was evaluated. These aircraft were all finished in silver colored aluminium dope and bore tail markings as described above. The registration was from the D-IAAA group. They were evaluated and operated by the DVL *Reklame-Staffel Mitteldeutschland* (Central German Publicity Squadron).

One other permanent marking should be mentioned at this point. At the rear of the fuselage, on the port side only, a legend was stencilled which gave some basic but essential information. This provided the name and address of the owner, the empty weight, the payload and maximum all-up weight, the maximum number of persons to be carried, including the crew, the date of last major inspection and, in some cases, the date of the next major inspection. The order and amount of detail listed varied somewhat but the size and positioning remained fairly constant. These details had to be applied in characters of at least 4-mm thickness and 25-mm in height in a dark color against a contrasting background. An example of a legend on a He 46 read as follows:

FLUGZEUGHALTER:TECHN.
SCHULE ADLERSHOF
(Owner)
BERLIN ADLERSHOF
(Address)
GRUPPE: LB1 H4
(Operator)
LEERGEWICHT: 1725KG
(Empty weight)
GESAMTLAST: 545KG
(Payload)
HÖCHSTZUL. FLUGGEWICHT: 2270KG
(Maximum all-up weight)
HÖCHSTZUL. PERSONENZAHL: 2
(Maximum number of persons)
LETZTE PRÜFG: 18.11.37
(Last major inspection)

Shortly after the public revelation of the *Luftwaffe* in March 1935, aircraft fin and rudder markings were revised and new machines entering service carried the stylised National Socialist flag marking across both sides. Implementation of this revision was slow at first and in many cases was not retrospective. Hence many aircraft were to be seen carrying the old-style combination tail markings up to a year or more later.

Civil registrations remained in use for *Luftwaffe* aircraft until the middle of 1936, at which time the first of a series of major changes to markings took place. A new military code system, comprising a five-figure/letter combination, was applied to all front-line operational aircraft. The proportions for these are shown in Diagram 3.

Being a military force, the *Luftwaffe* naturally identified its aircraft with a distinct series of markings. A *Balkenkreuz,* open-ended and of similar style to that employed during the closing weeks of World War I, was adopted and displayed on the fuselage and upper and lower wing surfaces. This black cross, with its narrow white border, was applied in constant size to all six positions in the proportions shown in Diagram 4.

On the fuselage, the first pair of numerals was always followed by the *Balkenkreuz,* then a single letter followed by two more numerals. This coding was repeated on the upper surface of the top wings and the lower surface of the bottom wings. These were positioned so that the first pair of digits appeared just inboard of the *Balkenkreuz* with the single letter on the centre line of the centre section and the remaining pair of digits just inboard of the starboard *Balkenkreuz.* The same layout was used on the bottom wings with the single letter on the centre line of the fuselage. (See Diagram 5). In the case of aircraft with a complex undercarriage system of struts — seaplanes particularly — to have placed this letter in the central position would have meant its being obscured from view. Therefore, in such cases, it was usually grouped with the last pair of numbers beneath the port wing. Applied in black, these codes contrasted strongly with the newly-introduced pale greenish-grey overall finish which was now gradually superseding the earlier silver finish.

To understand the code system it is first necessary to understand the basic structure of the *Luftwaffe* at this time. It was divided into six *Luftkreiskommandos* (Local Air Commands) with their individual HQ at Königsberg, Berlin, Dresden, Münster, München, and Kiel respectively, the latter being responsible for sea-based aircraft. Each *Luftkreiskommando* supported various *Geschwader* which were again divided into three *Gruppen,* these, in turn, being divided into three *Staffeln.* On paper the distinction was made by the use of Roman numerals for *Gruppen* and Arabic numerals for *Staffeln.*

The code group carried thus provided the following information:

1st figure — The *Luftkreis.*
2nd figure — The numerical sequence of the formation of the *Geschwader* within the *Luftkreis.*
Letter — The aircraft's individual code letter.
3rd figure — The *Gruppe* within the *Geschwader.*
4th figure — The *Staffel* within the *Gruppe.*

Thus the code 32 + A25 carried by a Do 23G of *Kampfgeschwader* 253 gave the following information:

3 — *Luftkreiskommando III,* Dresden.
2 — Second *Geschwader* of the *Luftkreis.*
A — Aircraft A within the *Staffel.*
2 — II *Gruppe.*
5 — 5 *Staffel.*

It will be noted that the coding did not compromise the security of the actual identity of the *Kampfgeschwader.*

Left. Detail photograph showing the Schwarz company emblem of a stylised black eagle on the propeller blades. The glossy pale grey coloring of the propeller blades is well shown. The spinner tip was painted in the main airframe color of pale greenish-grey. Right. The glossy finish of both the pale greenish-grey coloring and the military markings is quite apparent from this close-up shot of W Nr 1084.

While the presentation of the coding remained the same for seaplane units, a subtle variation was involved. Since the number of seaplanes was never sufficient to warrant a full *Geschwader* being established, the second digit was therefore superfluous. Instead of deleting it however, a naught was substituted. By reading the code in reverse, starting at the second to last digit and ignoring the code letter, the identity of the *Gruppe* was revealed. Thus the code 60 + U51 carried by a He 114 of 1/Kü F1 Gr 506 was interpreted as follows:

6 — *Luftkreiskommando VI,* Kiel.
0 —
U — Aircraft U within the *Staffel.*
5 — V *Gruppe.*
1 — 1 *Staffel.*

A further distinction applied to aircraft of the training schools, which also bore a five-digit/letter coding but with variations that are explained in Chapter 4.

Certain *Jagdgeschwader* were honored by the bestowal of traditional colors which were displayed prominently upon their aircraft. The actual areas of application varied somewhat from unit to unit and particularly between the two main aircraft types to carry them, the He 51 and the Ar 68, but the general area of the engine cowling was common to all.

The colors and the units to which they were allocated are listed below. Two of the units were also 'named', JG 132 *Richthofen* and JG 134 *Horst Wessel,* hence the respective allocations of the colors red and brown.

JG 131 based at Jesau — Black
JG 132 based at Döberitz — Red
JG 134 based at Dortmund — Brown
JG 232 based at Bernburg — Green
JG 233 based at Bad Aibling — Blue
JG 234 based at Köln — Orange

The first recorded 'variation' in *Luftwaffe* military markings occurred during the early days of the application of the new markings and codes. Photographic evidence so far limits this particular variation to I/JG 132 and I/JG 134 aircraft. Since these were the first of the *Luftwaffe's* new *Jagdgeschwader,* this may well

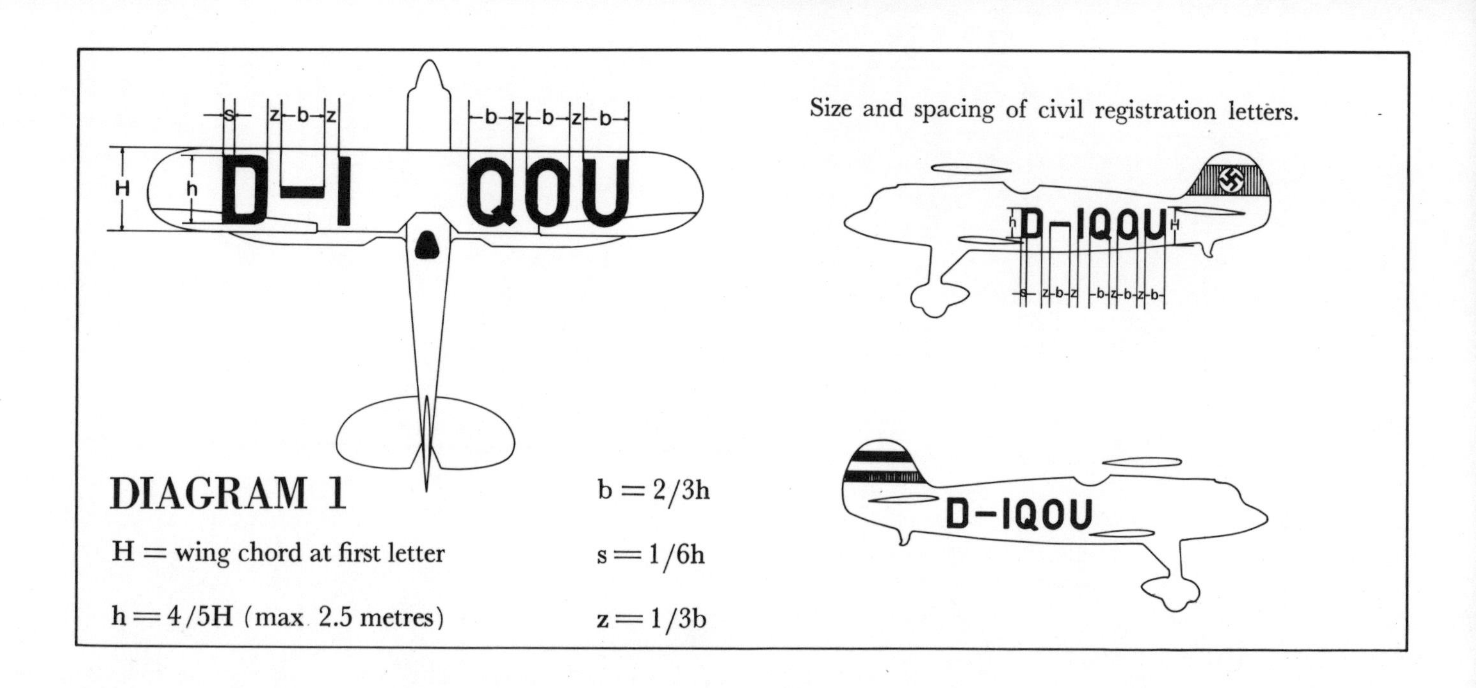

DIAGRAM 1

Size and spacing of civil registration letters.

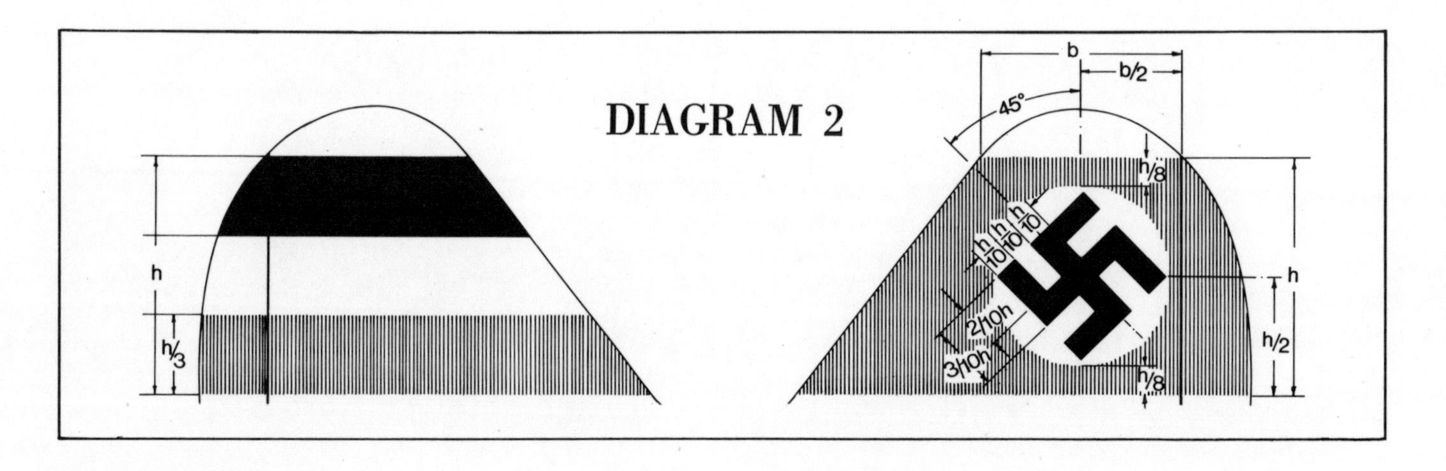

DIAGRAM 2

DIAGRAM 3

Proportions for aircraft letters and numbers.

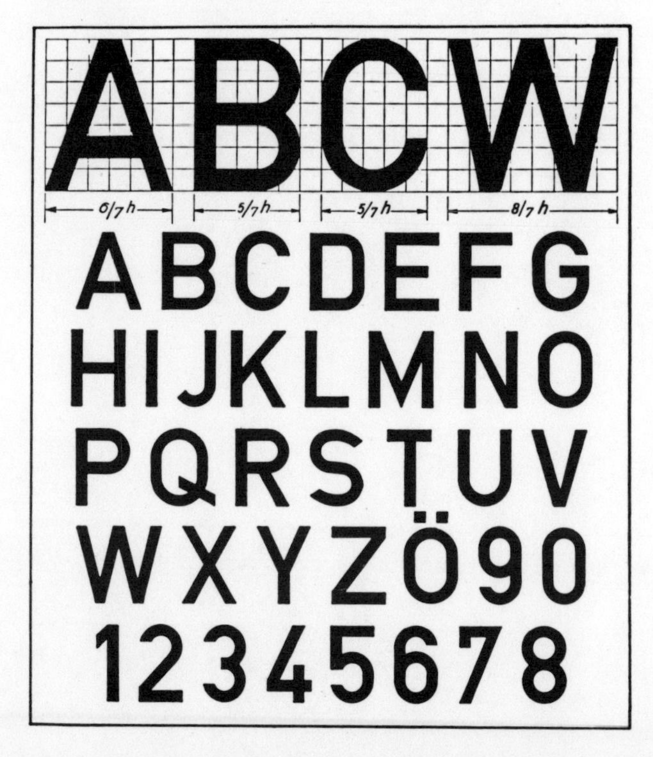

DIAGRAM 4

The earliest black and white *Balkenkreuz*.

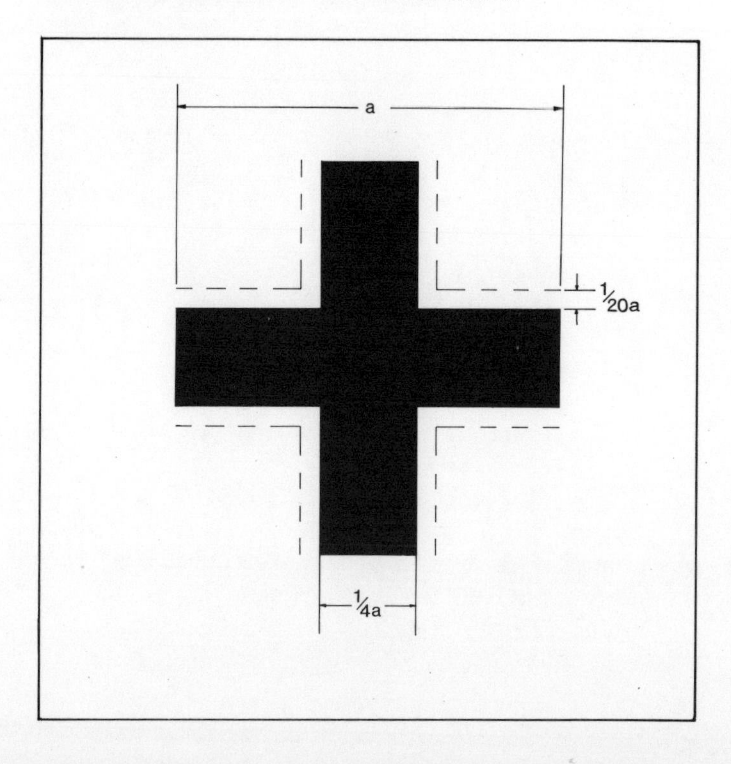

Ar 68Fs of JG 134 during the official presentation ceremony to the *Führer* on 15th May 1936. They bore full military markings but still had not had their traditional brown unit color applied to the engine cowlings. The overall color of the aircraft was pale greenish-grey, including the tip of each spinner. The propellers were finished with a dark wood stain. The exhaust stack cover plate was painted black to eliminate unsightly exhaust stains. A small white-edged, brown triangle on the lower engine cowl identified the oil filler point underneath. On the inside of the wheel spat was what appeared to be a black rectangle but which was, in fact, an instruction plaque. The dark trim area along the upper fuselage decking aft of the cockpit is visible on the second aircraft. This feature has often been mistaken for part of the traditional unit color markings.

Of poor quality but historically very interesting, this photograph of He 51s of I/JG 132 show the grossly oversized markings and codes. In the background is 21 + G11 which, like 21 + D11, initially still carried the starboard side fin and rudder markings after the introduction of military codes. The unit's traditional red color had been applied to the engine cowling area on both aircraft and even included the propeller blades and spinners.

Equally poor in quality but just as intriguing is this group shot of He 51s of JG 132 and JG 134. All nine aircraft carried the revised starboard side fin and rudder markings with the tri-colored striping replaced by the *Reichs-und Nationalflagge*. Commencing at the left, the first three aircraft appear to have been newly delivered since they still carried the white factory identification numbers, 55, 45 and 53. In the centre a pair of JG 134 aircraft flank a JG 132 aircraft. Like its companions on the right, the JG 132 aircraft had its nose section painted red and carried codes and markings of standard size, proportions, and positioning. The two JG 134 aircraft however bore the grossly oversized markings referred to earlier. The nearer of the two was coded 23 + Y12 but neither it nor its unit companion had the traditional brown color applied to the nose section. The lack of wheel spats on these 11/JG 132 aircraft was a unit modification.

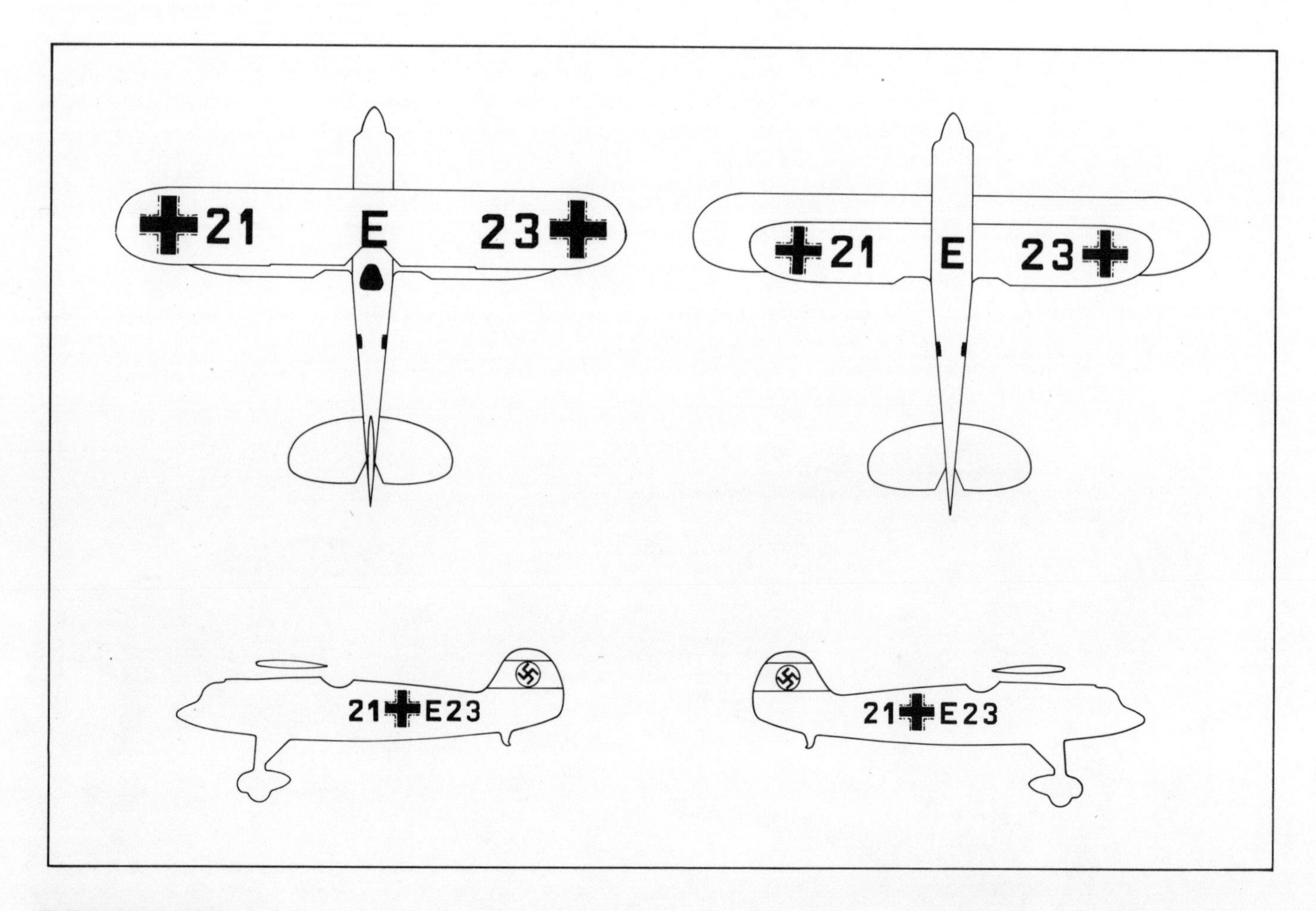

DIAGRAM 5

Location of codes for military aircraft. Aircraft with a complex undercarriage structure had the aircraft letter moved from the fuselage ventral centre line position and grouped with the two numerals beneath the port wing.

This closer view of aircraft of II/JG 132 shows the actual area of application of the red coloring. Each aircraft also bears the name of a species of hunting cat, Jaguar, Panther, etc, in white. Note that the second aircraft in the line-up had its entire spinner and propeller blades painted red. The weights and loading legend is just visible beneath the leading edge of the port side horizontal tail surfaces.

Two factory fresh He 51A-1s showing the style and size of the factory identification number applied to the fuselage sides in white. The *Balkenkreuz* still lacked a black edging to the white surround.

Prototype and experimental aircraft provided an individualistic series of markings variations. This was one of seven Hs 122B-0s and was photographed in mid-1936. At least three of the seven initially displayed only the fin and rudder markings seen here.

A group of Ju 52/3m g3e bombers of I/KG 152 *Hindenburg* wearing the newly introduced military codes. The *Balkenkreuze* still lacked any black edging to the white surround.

A tactical reconnaissance He 46C of Aufkl Gr (H)/122 showing the proportions of the code numbers to the *Balkenkreuz*. The lack of any black edging to the *Balkenkreuz* and codes helps to confine the photograph to the mid-1936 period. The numeral 2 signified that *Luftkreiskommando* II, Berlin, was the controlling authority but the use of 0 is interesting. It would appear that like the naval aviation *Gruppen* there were, at least pre-war, insufficient numbers to form a full *Geschwader*. Unlike their naval counterparts however, the identification of the *Gruppe* could not be determined from the actual elements of the code.

A long-range reconnaissance He 70F-1 of Aufkl Gr (F)/122 finished in overall pale greenish-grey. It retained the stylish Heinkel company trim. Again the *Balkenkreuze* still lacked a black edging.

Markings with a mystery. The first aircraft in the line-up, a He 51, bears the early-style military codes carried by fighter aircraft for a few months only. The code however, is intriguing. It begins with 0 and ends with 0 and appears to be 05 (or 3) + F (or E) 20. The significance of the first 0 is that it meant that the aircraft's unit was not part of a normal *Luftkreis*. The last 0 is equally interesting since it meant that the aircraft itself did not belong to a *Staffel*. The narrow black edging to the white surround on the *Balkenkreuz* indicates that the photograph was taken after mid-1936 but presumably before 1st September 1936. The He 50 and two He 45s were types then in front-line service so the use of civil registrations suggests that they were serving with a training establishment.

account for the irregularity in the application of the markings and codes. It is possibly also significant that photographs of aircraft from II/JG 132, formed only a couple of months later, show them carrying what became established as markings and codes of standard size.

The He 51s of I/JG 132 and I/JG 134 displayed markings of correct proportions but they were excessively large overall. The wing *Balkenkreuze* were of full chord, and those on the fuselage sides utilised virtually the entire available depth when viewed in side elevation. The codes were of correct proportion in relation to the markings but were therefore equally massive. The size of the codes in relation to the space available required the fuselage *Balkenkreuz* to be further aft on the starboard side than the one on the port side. Although space limitations were not so critical on the wing surfaces, the codes were again grouped most unusually, being separated by equal spacing, right across the wing between the two *Balkenkreuze*. The color illustration on Page 9 shows an example of this style of application.

Shortly after the introduction of the new markings, a further small revision was made to the *Balkenkreuz*. The narrow white border was outlined along its sides very thinly in black. (See Diagram 6). This was done to increase the visual identification range of the *Balkenkreuz* because of the low contrast between the white and the pale greenish-grey finish of the aircraft.

The adoption and use of the new code system was, at least for the *Jagdgruppen*, to be relatively brief as the *Reichsluftfahrtministerium* (German Air Ministry) pursued its policy of constant revision and improvement. On 2nd July 1936, directive LA Nr 1290/36 geh LA II/F1. In 3 was implemented and with it a radical change in fighter codes.

The new system, designed to aid the rapid air-to-air identification so necessary in fighter operations, reduced the actual code to a simple numerical sequence. Each *Staffel* utilised the numbers 1 to 12 which were applied in white and usually very thinly outlined in black. These numbers appeared in four positions; each side of the fuselage, on top of the upper wing centre section, and on the bottom of the fuselage between the wing roots. Dimensions were identical in all four positions.

A series of symbols was now used in conjunction with the *Staffel* numbers. This served to identify the *Staffel* within the *Gruppe* and the *Gruppe* within the *Geschwader*. (See Diagram 7 which explains the structure of a standard *Jagdgeschwader*). In addition a distinctive set of symbols was used to identify aircraft of the various staff flights. Official orders called for the introduction of all these markings by 1st September 1936.

Upper. A profusion of the revised style of fighter codes introduced in September 1936. The aircraft were from 1 *Staffel* of I *Gruppe* but the unit's identification is not known. Note that all the *Balkenkreuze* had narrow black edges to their white borders. The wing *Balkenkreuze* were located on the mean longitudinal centre line and thus overlapped the control surfaces. Among the aircraft in the background were three Fw 56s wearing the codes of *Küstenjagdstaffel* 136, the full code of one being 60 + W42. Lower. The white disc marking of 3 *Staffel* is clearly visible in this photograph. In the foreground is W Nr 2150, with W Nr 2155 to the left, while in the background D-IYJU is just visible. Note the high gloss finish on the upper surfaces of the bottom wings.

He 51B-1s of JG 134 on an airfield near Dortmund in the late summer of 1936. They display the white disc marking of 3 *Staffel.* The long-range fuel tank beneath the fuselage was painted in the same pale greenish-grey as the aircraft. Despite the late date the aircraft still do not bear their traditional brown unit coloring.

An Ar 68F undergoing maintenance shows its II *Gruppe* horizontal white bar marking.

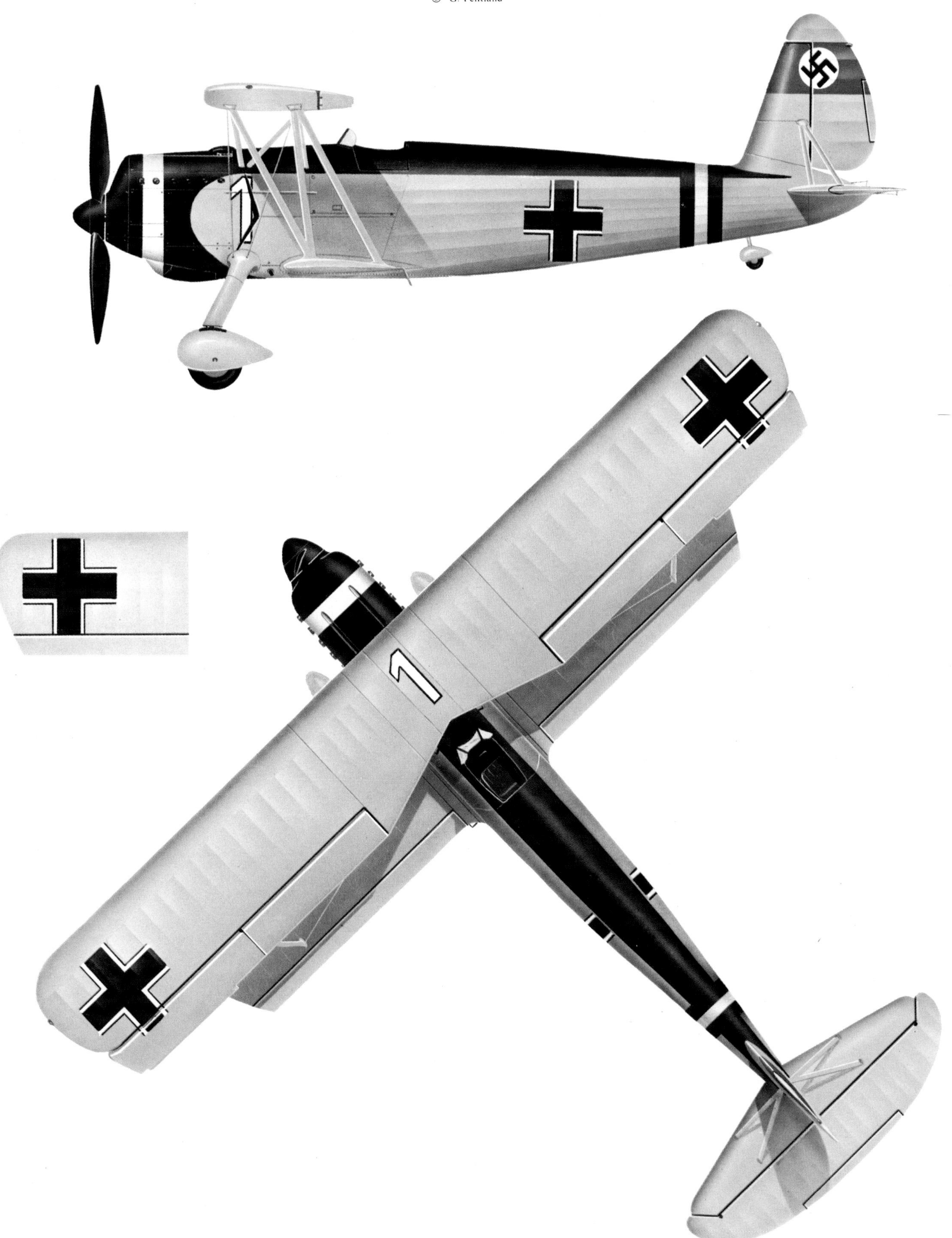

An Ar 68F of 2 /JG 131, based at Jesau, bearing the unit's traditional color of black. The white stripe around the engine and rear fuselage denoted 2 *Staffel* while the lack of any symbol aft of the number signified I *Gruppe*. The pale greenish-grey finish used by fighter aircraft required the *Staffel* stripe around the rear fuselage to be backed by a panel of contrasting color, where carried, or alternatively with the same dark blue used for the dorsal trim on many aircraft. The aircraft's number was repeated on the upper surface of the top wing centre section and between the wing roots on the underside of the fuselage.

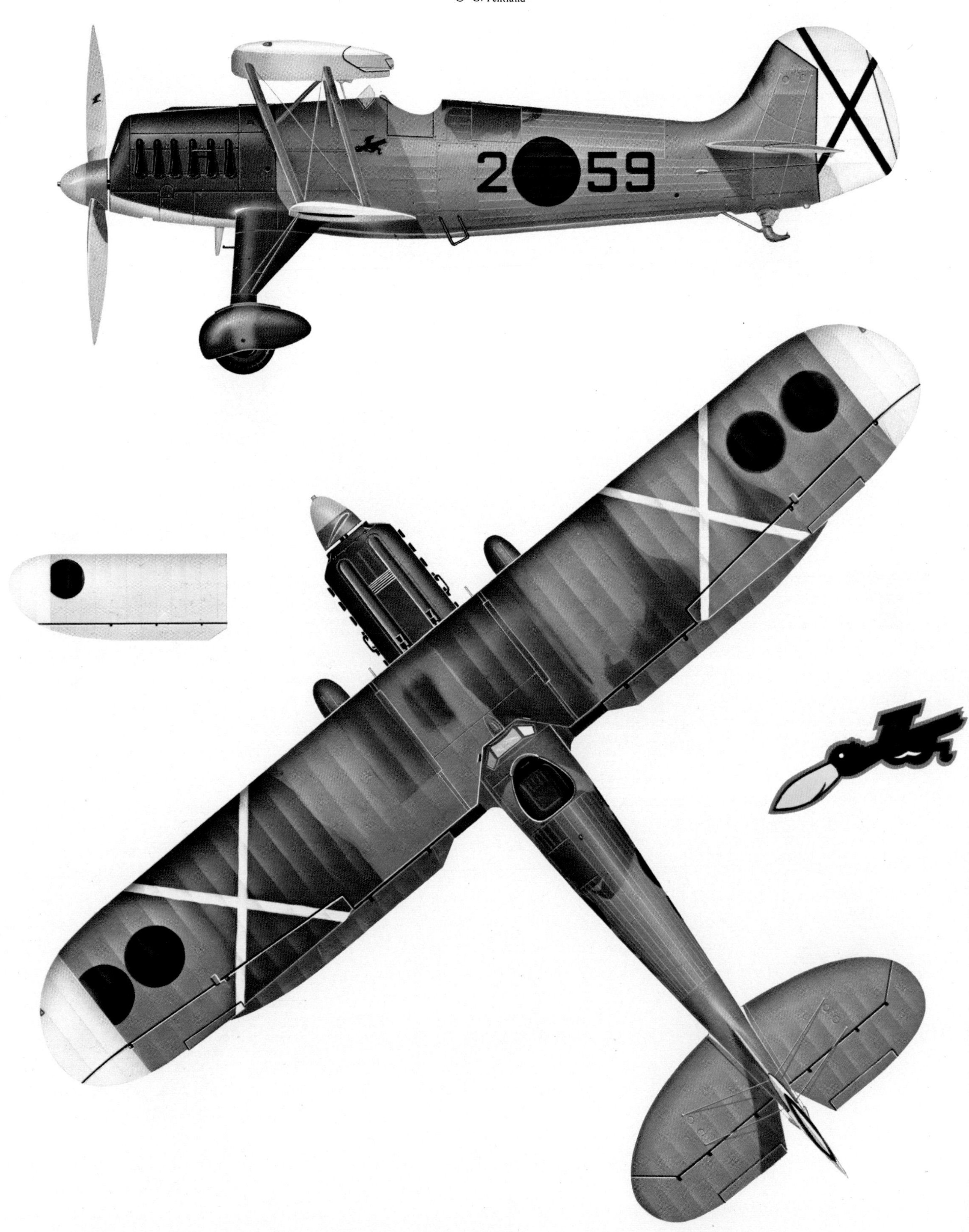

A good example of camouflage applied in the field. This He 51 of 1 J/88 shows the use of double black discs on the upper surfaces of the top wings. Note that those on the starboard wingtip were set further inboard. The hand painted white crosses were also rather crudely applied. The dark brown paintwork on the nose section occupied a similar area to that of the traditional color carried on some home-based aircraft. This may have been instigated by either the pilot or the ground crew, but the primary use of the color was undoubtedly for camouflage. The diving raven emblem appeared on both sides of the fuselage and was later adopted by St G 1.

© G. Pentland

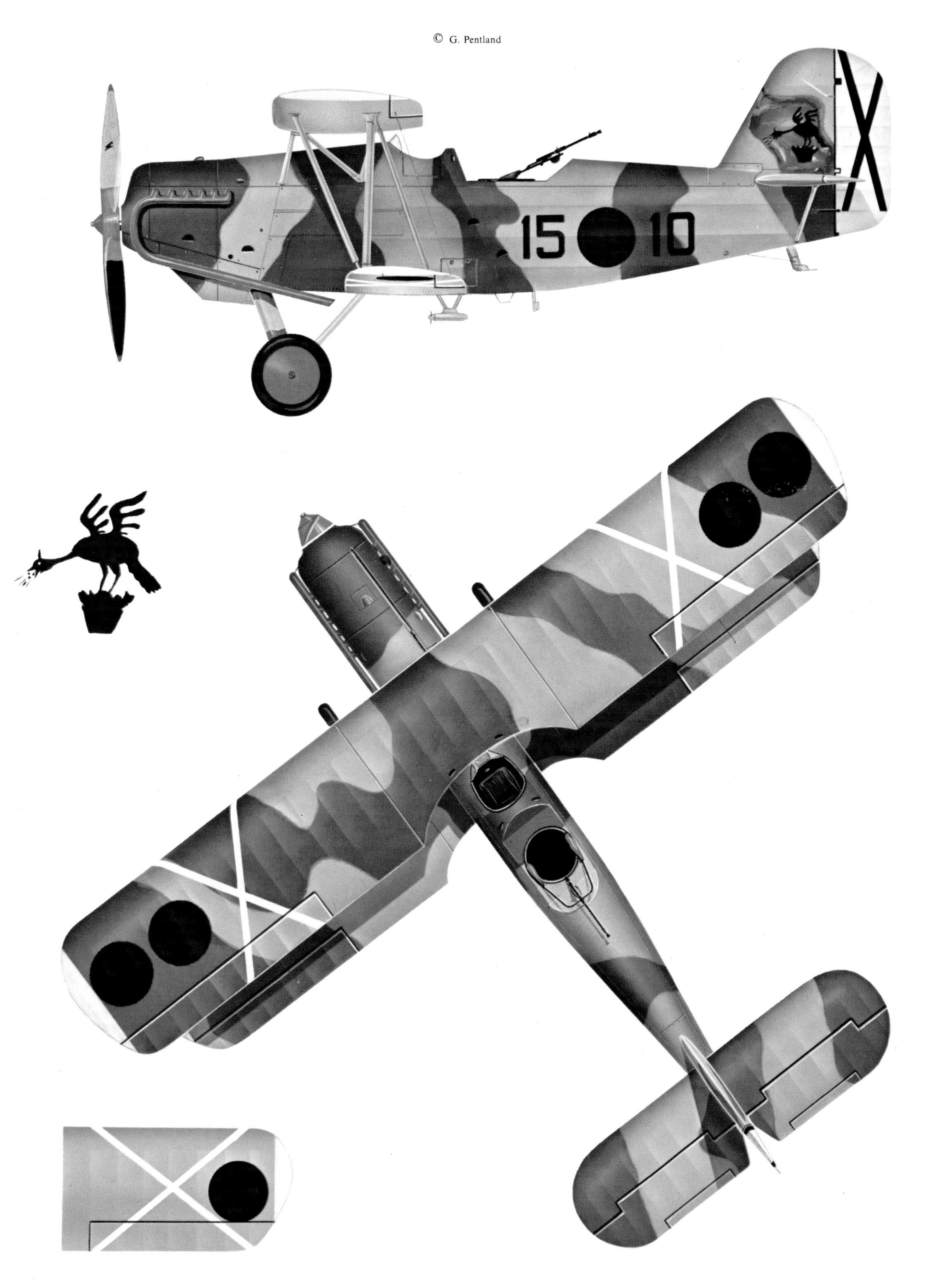

A more conventional form of camouflage was carried by this He 45C of A/88. The original pale greenish-grey finish had been oversprayed with continuous segments of dark brown. The upper surfaces of the top wings bore the double black discs, with those of the starboard wing again set a little further inboard. The white wingtip markings were narrower than usual on this particular aircraft. The fin emblem was possibly a personal marking.

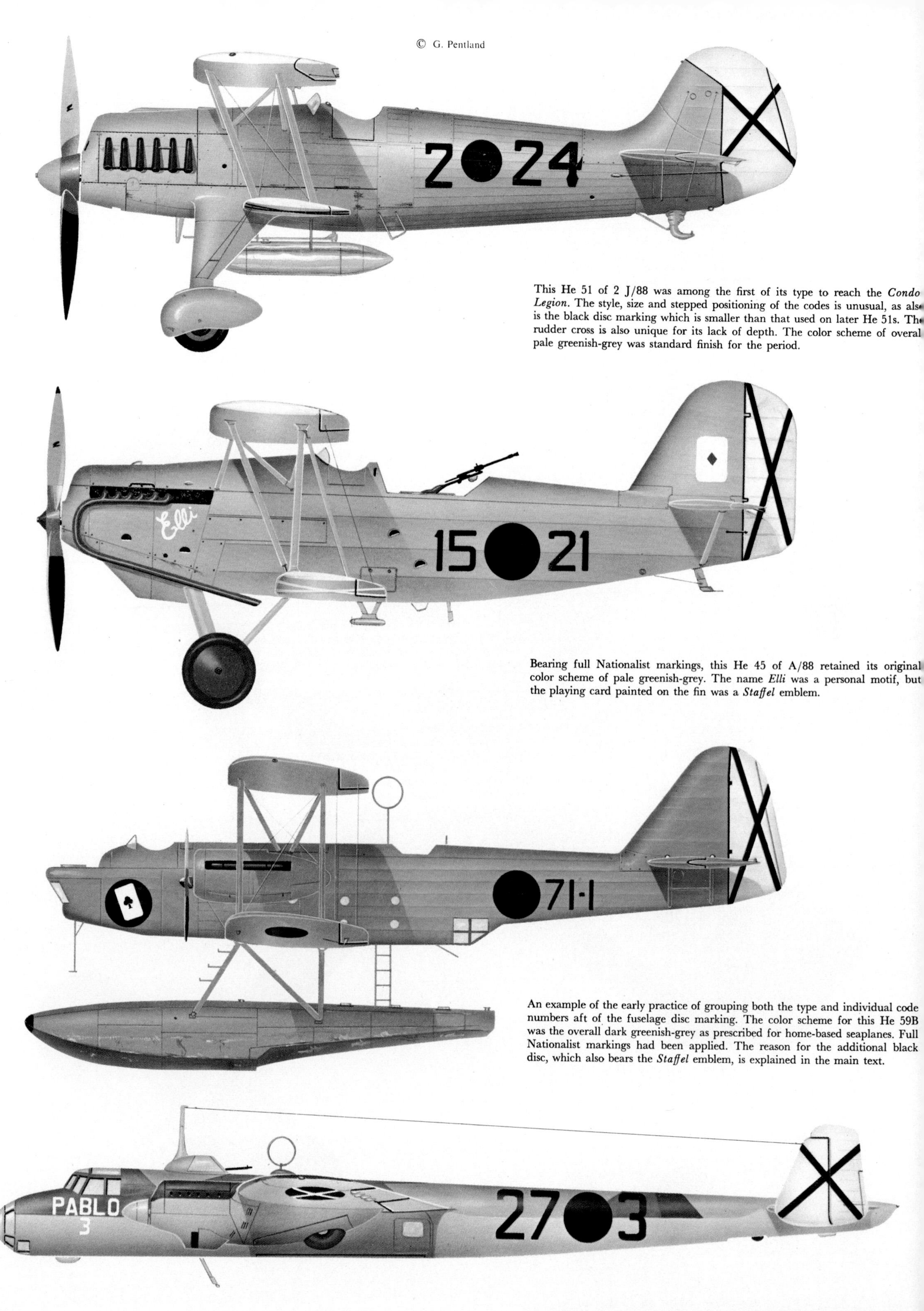

This He 51 of 2 J/88 was among the first of its type to reach the *Condor Legion*. The style, size and stepped positioning of the codes is unusual, as also is the black disc marking which is smaller than that used on later He 51s. The rudder cross is also unique for its lack of depth. The color scheme of overall pale greenish-grey was standard finish for the period.

Bearing full Nationalist markings, this He 45 of A/88 retained its original color scheme of pale greenish-grey. The name *Elli* was a personal motif, but the playing card painted on the fin was a *Staffel* emblem.

An example of the early practice of grouping both the type and individual code numbers aft of the fuselage disc marking. The color scheme for this He 59B was the overall dark greenish-grey as prescribed for home-based seaplanes. Full Nationalist markings had been applied. The reason for the additional black disc, which also bears the *Staffel* emblem, is explained in the main text.

Camouflaged in accordance with standards laid down for home-based bomber aircraft—dark brown, green, pale greenish-grey and blue grey—this Do 17 also carried the local nickname adopted for the type. Beneath the name appeared the aircraft's individual code number repeated in white.

The He 51B-2 was produced in small numbers as a seaplane version of this fighter and saw service with *Kürstenjagdgruppe* 136. The dark painted undersides of the floats were finished with a special anti-fouling compound.

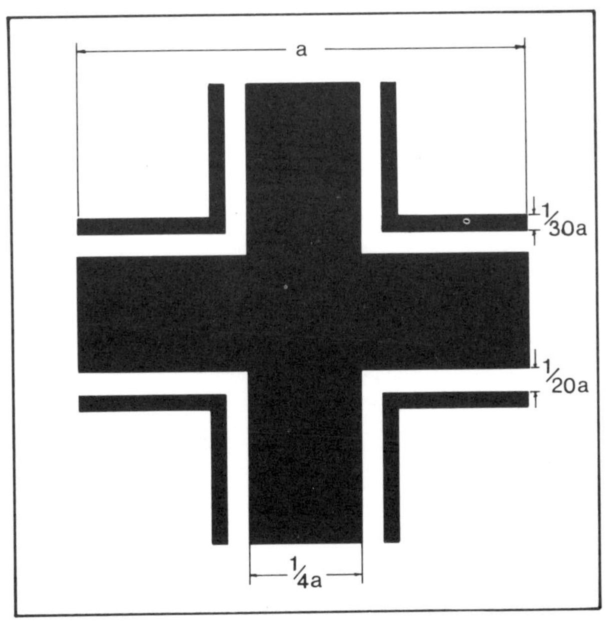

DIAGRAM 6

The revised *Balkenkreuz* with narrow white border outlined in black as mentioned on Page 24.

A Do 15 *Militär-Wal* 33 of 2 /Kü F Gr 106, formerly *Fliegerstaffel* (F) List, shows its codes from which it is possible to obtain the identity of the unit as detailed in the text. The undersurfaces were silver with all upper surfaces finished in dark greenish-grey. The propeller blades were natural metal but the tips were painted red, yellow and blue, the red being outermost.

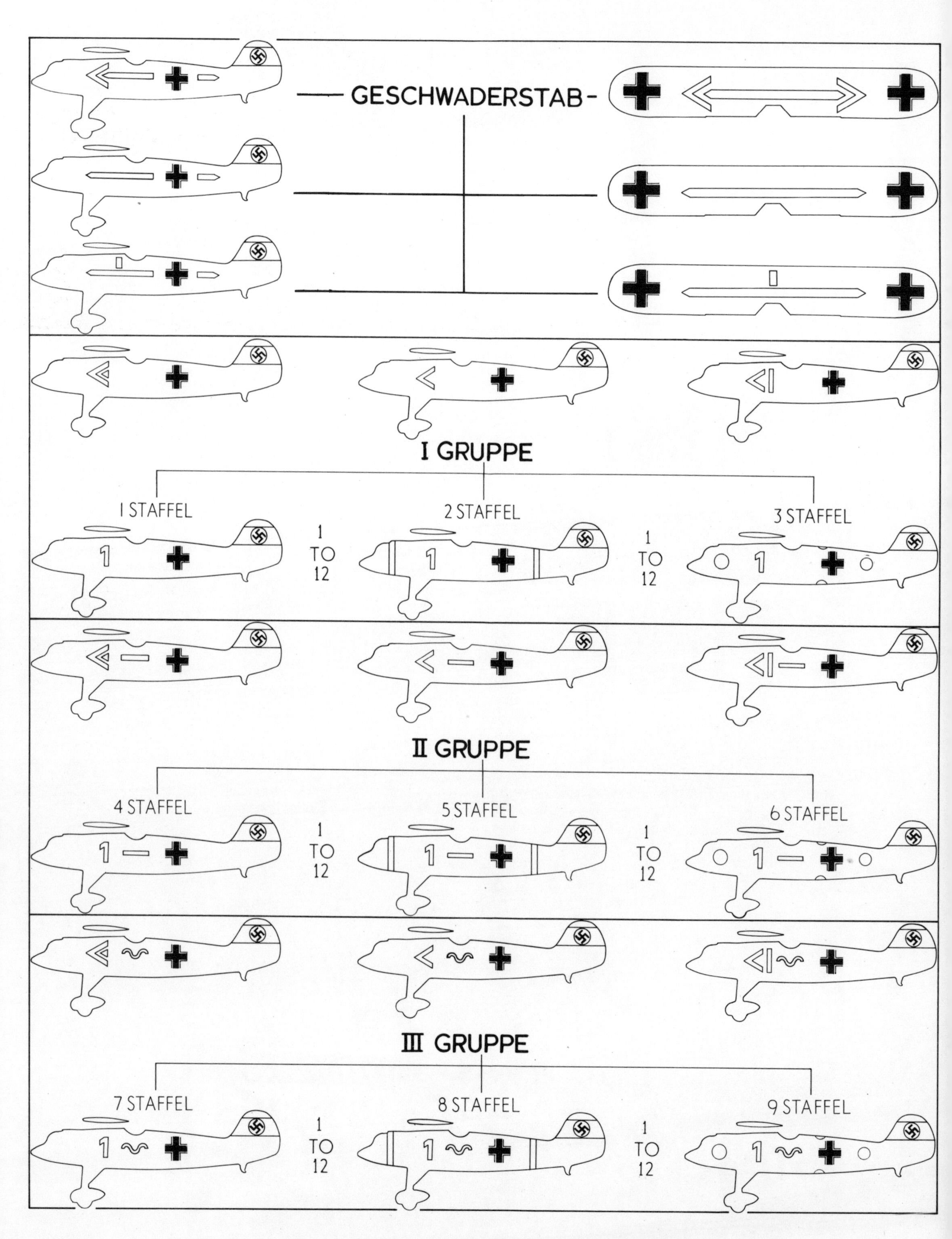

GESCHWADERSTAB
I GRUPPE
I STAFFEL
2 STAFFEL
3 STAFFEL
1
TO
12
1
TO
12
II GRUPPE
4 STAFFEL
5 STAFFEL
6 STAFFEL
1
TO
12
1
TO
12
III GRUPPE
7 STAFFEL
8 STAFFEL
9 STAFFEL
1
TO
12
1
TO
12

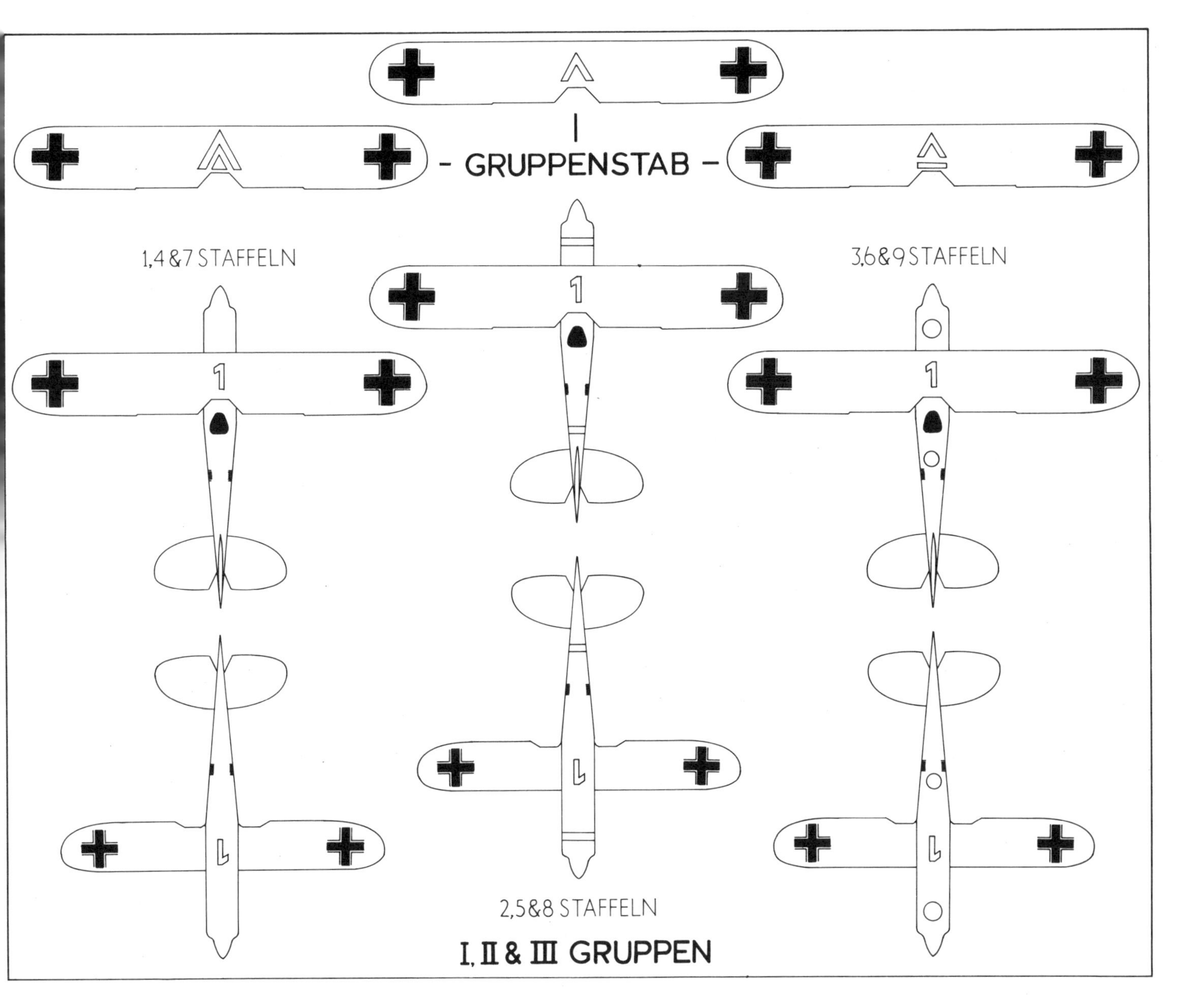

DIAGRAM 7

Geschwader, Gruppe and *Staffel* markings are shown on these two pages in schematic form. On the opposite page is depicted the basic structure of an entire *Geschwader,* while above is shown a *Geschwader.* Note that in both dorsal and ventral plan views the individual *Gruppe* identity could not be determined; only *Staffel* markings could be seen and even this information was limited. The quick reference table below condenses the main information shown on other diagrams.

GRUPPE	I	II	III	STAFFEL SYMBOL	
STAFFEL	1	4	7		
	2	5	8	WHITE BAND	Eng 300mm Fus 200mm
	3	6	9	WHITE DISC	Eng 500mm Fus 400mm
GRUPPE SYMBOL		1100mm	1098mm		

A Hs 123 coded 24 • 5, one of five that were sent to Spain in late December 1936. It carried the then current color scheme used by home-based ground support and bomber aircraft of dark brown, green, and pale greenish-grey with blue-grey undersurfaces. Full Spanish Nationalist markings had been applied, but unlike other biplanes there was one distinction. The narrow chord of the bottom wing and the restricted space available outboard of the bomb racks made it impracticable to include a separate white cross and a black disc marking. The two markings were thus combined in the same way that they were applied to monoplanes. (The same space limitations were the reason why home-based Hs 123s carried their underwing *Balkenkreuze* on the lower surface of the top wing.) A fire-breathing devil's head emblem, bearing the legend *Teufel* beneath it, was painted on the fuselage side just below the centre section struts.

CHAPTER 2

IBERIAN INTERLUDE

While the *Luftwaffe* was testing and improving its techniques and equipment, the tragedy of civil war broke out in Spain in July 1936. Despite much of what appeared in the press during the first year, the war was fundamentally a Spanish affair in origin and only after its outbreak did it achieve international significance. Many of the world powers sent aid of various forms to one side or the other, but Germany and Italy proved to be the major supporters of *Generalissimo* Franco's Nationalist forces.

Initial German aid consisted of six He 51 fighters and 20 of the versatile Ju 52s. They were all delivered in the usual overall color scheme of pale greenish-grey, the Ju 52s retaining their characteristic neat black trim around the engines and wheel spats. Heavy reversals in aerial supremacy however, caused a major increase in the aid given to Franco's forces by Germany, and a special force of some 200 aircraft was assembled and sent to Spain late in 1936. The *Legion Condor* had been born.

The force was divided into a bombing contingent of three *Staffeln* of Ju 52s designated *Kampfgruppe* 88 (K/88), a fighter contingent of three *Staffeln* of He 51s, designated *Jagdgruppe* 88 (J/88), a land-based reconnaissance contingent of 18 He 70s and six He 45s, designated *Aufklärungsstaffel* 88 (A/88), and a sea-based force of ten He 59 floatplanes for sea reconnaissance, designated *See-Aufklärungsstaffel* 88 (AS/88).

When the Spanish conflict erupted, a distinctive set of markings for each of the combatants became a vital necessity. Franco's Nationalists had originally intended to retain the pre-1931 colors of red/yellow/red but the obvious confusion with the red/yellow/purple of the Republican markings caused the Nationalists to adopt a simpler method of identification. Thus rudders were painted white with a black diagonal cross, while wings and fuselages bore a solid black disc.

To conform to international law the *Legion Condor's* aircraft adopted Nationalist markings. In addition, a simple numerical code was introduced, being applied in solid black figures to the fuselage sides only. Each type of aircraft had its identifying code, ie 2 for the He 51, 14 for the He 70, 15 for the He 45, 22 for the Ju 52 etc. This code was generally applied to the left of the fuselage black disc marking with the aircraft's own code number to the right. The code was also a simple numerical sequence which, for security reasons, usually had a random starting point. All four types of aircraft retained their overall pale greenish-grey finish.

He 59 floatplanes were also delivered in standard finish for home-based marine aircraft, this also being greenish-grey but of a distinctly darker shade than that used by the land-based aircraft types of the *Legion Condor*. The He 59s also differed from the general style of code application, all the numerals being grouped aft of the rear fuselage black disc and separated by means of a hyphen. An additional black disc was painted on the forward fuselage beneath the cockpit. This extra disc was necessary since the bulk of the wing and engine complex obscured the rear fuselage markings from all angles other than the beam and the rear quarter.

Shortly after operations began, the difficulty of rapid air-to-air identification produced a revision of the markings applied to both the Nationalist and *Legion Condor* aircraft. The latter introduced a second black disc marking of identical proportions on each upper wingtip of the He 45s, He 51s and He 59s. Monoplanes were in a distinct minority and apparently the distinctive shapes and sizes of Ju 52s and He 70s were considered sufficient recognition features. The oceanic environment and nocturnal habits of the He 59s also eliminated identification problems where they were concerned.

The additional black disc marking however, proved ineffective and in common with the Nationalist air arm, white wingtips and full-chord white crosses were therefore hand painted on the wings. The full-chord white crosses were applied only to the existing biplane types, ie the He 45s, He 51s and He 59s, albeit with some distinct variations in size and proportion. Other aircraft types also had the white tip markings applied but the wing crosses were confined within the periphery of the existing black discs.

A special test force of five Hs 123s arrived in Spain in late December 1936 and the revised markings were also applied to them, but in modified form. Top wing upper surfaces had full chord white crosses but the bottom wing undersurfaces had the white crosses enclosed within black discs. The color scheme of these Hs 123s was the latest type then being introduced for

A He 51B-1 of 1 *Staffel* of J/88 at Leon in the summer of 1937 displays its field-modified camouflage scheme. All the upper surfaces and the fuselage sides had been sprayed with random patches of dark brown and green. The base color of pale greenish-grey showed through in places and a section of the fuselage underside, from just aft of the footrest to the tailskid brace, was also left in this color. The remainder of the undersurfaces had been sprayed blue-grey. On the original photograph it was possible to see the black disc marking on the undersurfaces of the bottom wing, between the white wingtip and the white cross. The spinner tip was also white.

Silhouetted against the sky by the early morning sun, the bold black and white rudder marking of 24 • 2, in the foreground, shows its value as a recognition feature.

bomber aircraft in Germany. All upper and side surfaces were finished in a segmented pattern using the three colors dark brown, green and the existing pale greenish-grey, with the undersurfaces in blue-grey. (See Chapter 3 for precise details of these colors.) Deliveries of replacement bomber aircraft and some new types during the next few months were also to exhibit this new camouflage scheme.

Kampfgruppe 88 received a boost early in the new year when it was decided to re-equip two *Staffeln* with the new He 111B. The first of these arrived in February 1937 and began operations in March. The earliest deliveries of the type still retained the overall pale greenish-grey finish to which the Nationalist markings were added. Like the existing monoplane equipment, He IIIs also had the wing crosses contained within the black disc markings.

Approximately ten of the first 20 aircraft had the home-based style of numbers used for their code application before adopting the more curved, flowing style in use on the *Legion Condor* aircraft. Later deliveries of the He IIIB-2 model arrived wearing the new-style bomber camouflage of dark brown, green, greenish-grey and blue-grey. These aircraft had their codes applied in the local manner. Other aircraft supplied to the *Legion Condor* over the next year or so were also to exhibit this method of identification.

The land-based reconnaissance unit also received fresh equipment with the arrival of 15 Do 17Fs in the spring of 1937. These were finished in the new bomber camouflage to which Nationalist markings were added. Due to the twin fin/rudder configuration, the tail markings were applied to both the inside and outside faces. In addition, the markings were applied right across both fin and rudder, the inside marking being somewhat abreviated due to the position of the tailplane. Like the early He 111s, the first two or three Do 17s used the more squared style of code numbers.

This simple functional style was used by most of the other types supplied to the *Legion Condor* during the early stages of the war. He 45s, He 51s, He 59s and Ju 52s retained this plain style of numbering, with some very minor variations, throughout the Spanish campaign.

The He 45s and He 51s were meanwhile subjected to some very original camouflage schemes. Irregular patches of the new dark brown paint were sprayed over the original overall greenish-grey color. In some instances the application took the form of small patches only but others consisted of continuous segments. Some

Of historical interest despite the poor quality, this photograph shows some of the first Bf 109B-1s to reach Spain. The second aircraft from the left was coded 6 • 16 and like other early arrivals had its code grouped aft of the fuselage black disc marking. Equally interesting was the lack of a small white cross on the fuselage black disc, especially since the underwing black discs *did* carry the usual white cross. The overall finish appears to be the standard pale greenish-grey applicable to home-based fighter aircraft. In the background was a Ju 86D coded 26 • 1. It bore the full four-color camouflage and had the name FUMO, in white block letters, beneath the cockpit with the number one, also in white, beneath that again. The presence of this aircraft dates the photograph as no earlier than 1937.

of the He 51s were more extensively renovated and were sprayed with a base color of the new camouflage green, over which the brown was sprayed in random patches. Undersurfaces either remained the original pale greenish-grey or were repainted with the new blue-grey color. Most applications, particularly the continuous segments and patches, appear to have been definitely sprayed on but a few repaint jobs to the lower surfaces, in blue-grey, had every appearance of being hand painted.

Jagdgruppe 88 also received some new equipment in April 1937 when the first Bf 109Bs arrived. Three prototypes had spent seven weeks on field testing in Spain before being shipped back to Germany, but actual production models were issued to the front-line unit and replaced the He 51s of 2 J/88.

Nationalist markings were applied to these early Bf 109s which had also arrived wearing the overall pale greenish-grey finish. Like the other monoplane types, wing crosses were confined within the black disc markings. In addition, the fuselage disc marking also bore a small white cross, but its arms did not extend to the circumference. This was a variation from the normal which was to be applied with some consistency and exclusively to the Bf 109. The remaining markings to wingtips and rudder conformed to the prevailing standard.

Like other early deliveries to Spain, these Bf 109s bore identification codes in plain functional style and in the earliest instances, with the entire code aft of the fuselage disc marking, the actual elements being separated by a hyphen. The Bf 109s however, soon adopted the more stylish form of numbering and it was because of the increased size of these that the black disc fuselage markings were not diametrically opposite each other, the port side one being slightly further forward.

Like their fellow combatants, some of these early Bf 109s became the subject of field camouflage experiments. A few were merely given blue-grey undersurfaces but more radical alterations were to come. The Spanish terrain in summer is dry and parched and an overall application of dark brown paint was therefore considered suitable camouflage. Most of the Bf 109s so finished appear to have retained their original greenish-grey coloring for the undersurfaces. Unlike the He 45s and He 51s, the Bf 109s appear to have generally used an overall finish for their upper surfaces rather than a patch pattern of two colors. Spinners were usually repainted to match the main color but a few either retained their greenish-grey color or were repainted white, possibly to distinguish flight leaders.

Painting the exhaust areas on the engine cowlings made the removal of exhaust deposits easier and 2 J/88 adopted this practice, turning the black paint application into a form of stylish trim. The practice was continued until or if the aircraft were later repainted a darker camouflage color.

The Spanish theatre of operations also saw the extensive revival of both unit and personal emblems. The latter ranged from stylised letters of 1914-18 practice to cartoon-style characters. Indeed some of the Do 17s displayed some very original artwork as if to prove that

Another early Bf 109B of J/88. It was finished in overall pale greenish-grey and bore its codes in the more stylish local fashion.

Among the other early Bf 109Bs of J/88 was this one, 6 • 4?. It displayed a camouflage scheme similar to that shown in the color illustration of 6 • 51 on Page 47. Note the marked contrast of the dark brown of the wing upper surfaces with the green of the fuselage. The pale undersurface greenish-grey is just visible at the base of the fuselage.

Finished in overall pale greenish-grey, this He 111B was among the first batch of the type to reach Spain. The very original fin marking records the crew's devotion to their mascot who was killed during the air fighting over Sagunta. The date helps to illustrate the point that, unlike some of the fighter aircraft, a few of the early bomber aircraft delivered were not retrospectively camouflaged.

The other side of 25 • 15 showing not only the repetition of the fin illustration and legend (albeit that the top line had been applied by a less professional hand) but also the *Staffel* emblem. This emblem was later to be adopted by I/KG 53.

Another He 111B, 25 • 4, but this time wearing three-color upper surface camouflage. Compare the style of the code, and its positioning in relation to the black disc marking, with that of 25 • 15. The white cross overlapping the underwing black disc marking was unusual.

A He 111B at dispersal showing its multi-color camouflage. The *Staffel* emblem on the nose was a variation on the K/88 winged bomb emblem and, in a modified form, it was later adopted by 4 /KG 53. Compare the underwing marking with that shown on 25 • 4.

The first of two Ar 68Es sent to Spain where they were used for night-fighter trials. Finished in overall pale greenish-grey, it bore the Nationalist markings but the under-wing white crosses did not cover the full chord of the wing. The aircraft type code is not positively known but was either 8 or 9. The aircraft's own number was 1.

A rare photographic subject. One of the five Ju 86D-1s used by the *Legion Condor,* 26 • 3 was finished in the dark brown, green, pale greenish-grey and blue-grey scheme. Note that the Nationalist rudder markings were repeated on the inside face of the rudders. On the original print the name FUMO could be discerned in white block letters, with the number three, also in white, beneath it, and just visible below the pilot's cockpit. It was the practice of some of the Do 17, He 111 and Ju 86 sections to use the names Pablo, Pedro and Fumo, respectively, painted in white on the nose section with the aircraft's individual number painted in white beneath the name.

A He 111E, 15 • 63, from the batch delivered in 1938. The camouflage was of the standard variety. The emblem was later to be adopted by III/KG 53.

Typical of the late deliveries to the *Legion Condor* was this Fi 156. It carried the squarer-style coding applicable to home-based aircraft in Germany. The color scheme was also the later type of black-green and dark green with blue-grey undersurfaces.

such flamboyance was not the prerogative of the fighter units! The four *Staffeln* of J/88 adopted a diving raven for the first *Staffel,* a top hat for the second, a cartoon-style mouse for the third and an Ace of Spades for the fourth. The diving raven emblem was eventually replaced by a caricature of a head with a 'wooden' eye. Playing card emblems appear to have been a popular source of inspiration and variations appeared on both the He 45s and He 59s.

Another small but significant marking made its appearance in Spain. Small slender white bars were used to signify victory scores for the fighters and these were applied to the fin, near the tip. For some pilots, Spain was to see the start of a long line of such markings.

Several other new types of aircraft saw action with the *Legion Condor.* Five Ju 86s and three Ju 87As arrived in late 1937. In both cases they all wore the new home-style three-color segmented bomber camouflage to which was added the Nationalist markings. In October however, three Ju 87Bs arrived in company with six Hs 126s, the latter as replacements for the He 45s of A/88. Both types now wore a new style and color of upper surface camouflage consisting of black-green and dark green with blue-grey undersurfaces. (See Chapter 3 for details of these colors and the corresponding pattern).

Ten Do 17Ps which arrived to supplement the older Do 17Fs also wore this new camouflage scheme. Markings and code styles were applied as for the older aircraft with one variation; one or two of the new Do 17s had the Nationalist tail marking restricted to the rudder alone, although it was still applied to both sides.

With the appearance of the new color scheme, no further variations are known to have occurred. Later deliveries of aircraft all tended to have their code numbers applied in the angular, home-based style. It should perhaps be mentioned here that the code system used was fully integrated with that in use by the Nationalist forces in order to eliminate duplication and subsequent confusion. Of the many different machines used by the *Legion Condor* in Spain, the following is a list of the code allocations used to identify the various aircraft types.

Dornier	Do 17	—27
Fiesler	Fi 156	—46
Heinkel	He 45	—15
	He 51	— 2
	He 59	—71
	He 70	—14
	He 111	—25
Henschel	Hs 123	—24
	Hs 126	—19
Junkers	Ju 52	—22
	Ju 86	—26
	Ju 87	—29
	W 34	—43
Messerschmitt	Bf 108	—44
	Bf 109	— 6

Representative code allocations for some of the Spanish Nationalist operated aircraft are as follows:

Aero 100	—17
Bregeut 19	—10
Farman 190	—30
Heinkel He 112	— 5
Vultee V-1A	—18

Among the last aircraft to reach the German forces in Spain in time to see active combat service was a batch of Bf 109Es. These appear to have been finished with green upper surfaces and blue-grey undersurfaces. The exhaust areas were painted black to eliminate unsightly staining on the relatively light-colored sides. The second aircraft in the line-up was 6 • 115.

Photographed at Zurich in July 1937, the Do 17 V8, alias the Do 17M V1, wore the overall 63 finish common to prototype aircraft of that period. The application of its national and military markings was strictly in accordance with current *RLM* regulations. Another practice common to prototype aircraft was the application of the manufacturer's name on the forward fuselage.

Also photographed in mid-1937, the Hs 123 V5 displayed the consistency and style of markings applicable to prototypes. The unusual sesquiplane layout and narrow chord of the bottom wings resulted in the *Balkenkreuze* being marked beneath the main wings.

CHAPTER 3

PRELUDE TO WAR

While the Spanish conflict pursued its tragic course of fratricide, a steady change took place in *Luftwaffe* equipment, camouflage, and markings.

Aircraft paint finishes were carefully controlled by the *RLM* who, as in all other aspects of their administration, approached the subject with thoroughness. Comprehensive schedules were laid down to positively identify each and every paint type and color in use. To facilitate this, a system was introduced whereby each aircraft paint group *(Flieglackkette)* was identified by a two-figure code number. Groups 01 to 19 were used for metal surfaces only while groups 20 to 39 were used for wood and fabric surfaces. Individual aircraft paint types *(Flieglack)* were identified by a four-figure code number while the actual color pigment *(Farbton)* was identified by a two-figure code number written as a suffix, eg 7105.02.

The *Farbton* code used a number sequence commencing at 00 and ending, for the time period covered in this first volume, at 73. At first sight there would appear to be grounds for confusion between the paint group codes and the color pigment codes, at least up to the number 39. The technical directives issued however, always prefixed any of the 01 to 39 code numbers with the identification *Flieglackkette Nr*. The 00 to 73 code numbers were normally seen in directives only as a suffix to an actual paint type code, the latter being clearly identified with the title *Flieglack*.

The *Flieglackketten,* in addition to being split into two distinct groups, also had certain restrictions stipulated for each individual group as detailed below. Only some of the identification codes were used, the remainder presumably being reserved for allocation to latter paint developments.

Flieglackkette Nr 01 — This group consisted of *Flieglack* 7100, 7105, 7106, 7107 and 7240 plus paint thinner 7200. *Farbton* 02 was used in conjunction with 7105 and 7107. These paints were for use on all external metal parts with the exception of seaplane floats and the underwater sections of flying boat hulls. (This group was eventually phased out of use by 1941, being replaced by *Flieglackkette Nr 04.*) Note that *Farbton* 00, which was specified for use with *Flieglackverdünnung* 7200 was a *clear* finish.

Flieglackkette Nr 02 — This group consisted of *Flieglack* 7102, 7106, 7107, 7108 and 7200. Up to 1939, 7107 and 7108 were used respectively, in conjunction with *Farbtöne* 02 and 01, 00 being used with 7200. Application of these paints was restricted to the interior of metal seaplane floats, the interior and exterior of float struts and the interior of flying boat fuselages. (This group was modified in 1941.)

Flieglackkette Nr 03 — This group consisted of *Flieglack* 7100, 7105, 7106, 7107, 72400 and 7200. *Farbton* 01 was used in conjunction with 7105 and 7107, 00 being used with 7200. Uses and restrictions applicable to this group were the same as those for *Flieglackkette Nr 01.*

Flieglackkette Nr 04 — (The 1939 edition of *Der Fleugzeug-Maler* did not list this group as it was not introduced until after the outbreak of war.) This group consisted of *Flieglack* 7102, 7109 and 7200. *Farbton* 02 was used in conjunction with 7109 and any of the other standard camouflage *Farbtöne,* while 00 was used with 7200.

Flieglackkette Nr 05 — This group consisted of *Flieglack* 7113, 7114, 7115 and paint thinner 7213. *Farbton* 02 was used in conjunction with 7115 and any of the other standard camouflage *Farbtöne,* while 00 was used with 7213. All paints in this group were non-inflammable as their use was specifically for metal parts on shipborne aircraft. (Like *Flieglackkette Nr 04* this group was not introduced until after the outbreak of war.)

Flieglackkette Nr 20 — This group consisted of *Flieglack* 7130, 7135, 7136 and paint thinner 7230. *Farbton* 02 was used in conjunction with 7135, 00 being used with 7136. *Flieglack* 7130 was also used to bond fabric together, all paints in this group being restricted to use on fabric coverings. (After the outbreak of war 7135 was also listed for use in conjunction with any of the standard camouflage *Farbtöne.*)

Flieglackkette Nr 21 — This group consisted of *Flieglack* 7120, 7130, 7135 and 7230. *Farbton* 01 was used in conjunction with with 7135 and 00 with 7230. Paints in this group were restricted to external fabric finishes. (This group was phased out of use after the outbreak of war.)

Flieglackkette Nr 22 — (The 1939 edition of *Der Fleugzeug-Maler* did not list this group as it was introduced after the outbreak of war.) This group consisted of *Flieglack* 7115, 7137 and paint thinner 7213, plus paint spreader 7215. *Farbton* 02 was used in conjunction with 7115, while 00 was used with both 7213 and 7215. In addition, 7215 could be used in conjuction with any of the other standard camouflage *Farbtöne.* All paints in this group were non-inflammable as their use was specifically for fabric coverings on shipborne aircraft. Furthermore these paints could not be applied over the top of any other types of paints nor were they interchangeable with those of *Flieglackkette Nr 20.*

Flieglackkette Nr 30 — This group consisted of *Flieglack* 7131, 7132, 7135, 7136 and 7230. *Farbton* 02 was used in conjunction with 7135 and 00 with 7131, 7136

Two shots of the twelfth production Ju 87A-1 which was camouflaged in 61/62/63/65 and bore the class B1 registration D-IEAU. On delivery to a front-line unit, the black civil registration was removed and replaced by a military code. The number 12, in small white numerals beneath the tailplane, denotes the aircraft's production sequence. Upper surface camouflage colors were taken right down the fuselage sides and included the undercarriage spats on most early Ju 87s.

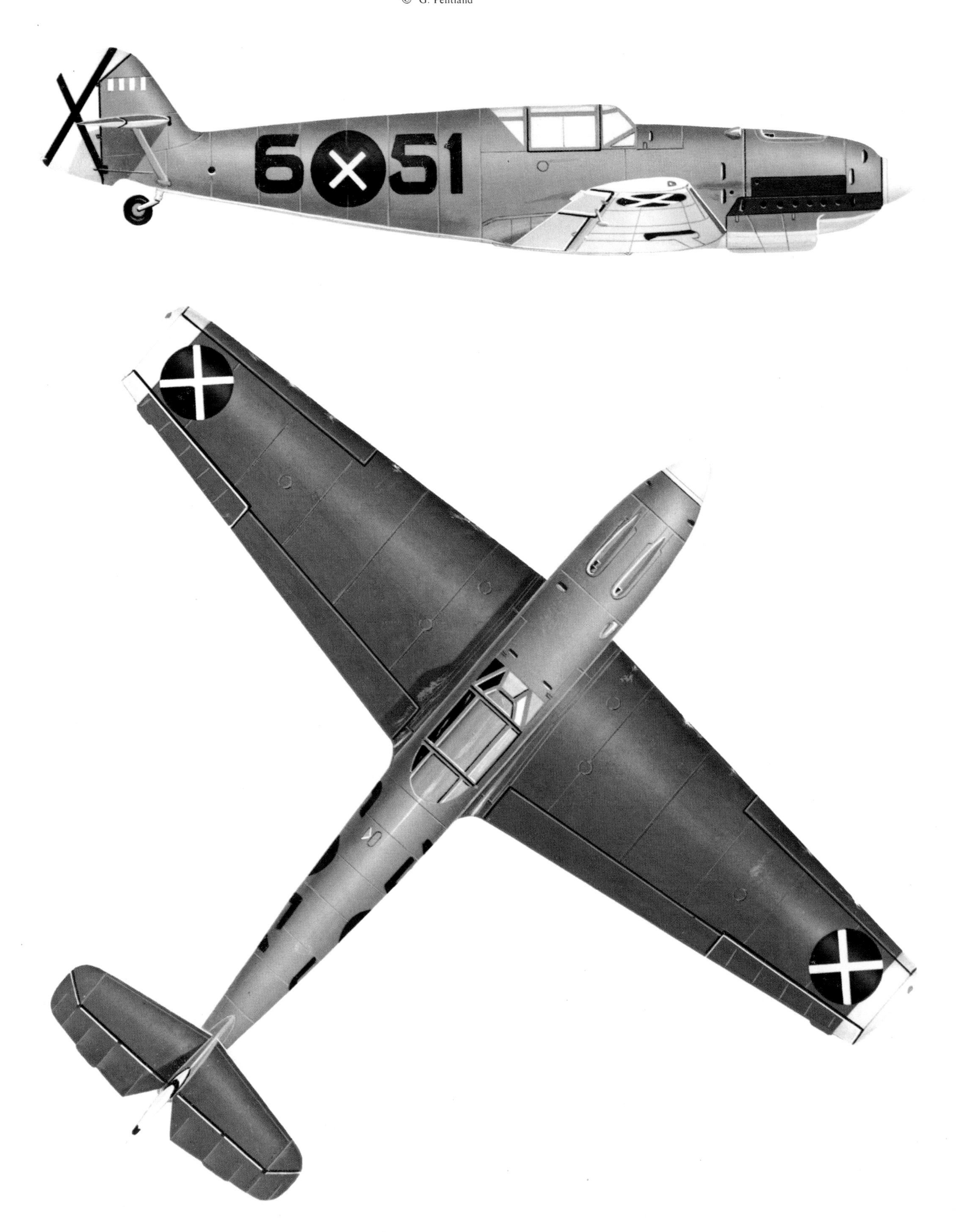

In general the Bf 109s of J/88 were repainted with a single upper surface color of either dark brown or green. At least three examples however have been recorded wearing the camouflage scheme seen above, although the undersurface color varied from the old pale greenish-grey to the newer blue-grey. (See also the photograph on Page 38.) The four kill markings were somewhat broader than examples seen later in the campaign. The code style was also the more flowing type which, because of space limitations, resulted in the black disc markings not being diametrically opposite each other on the fuselage.

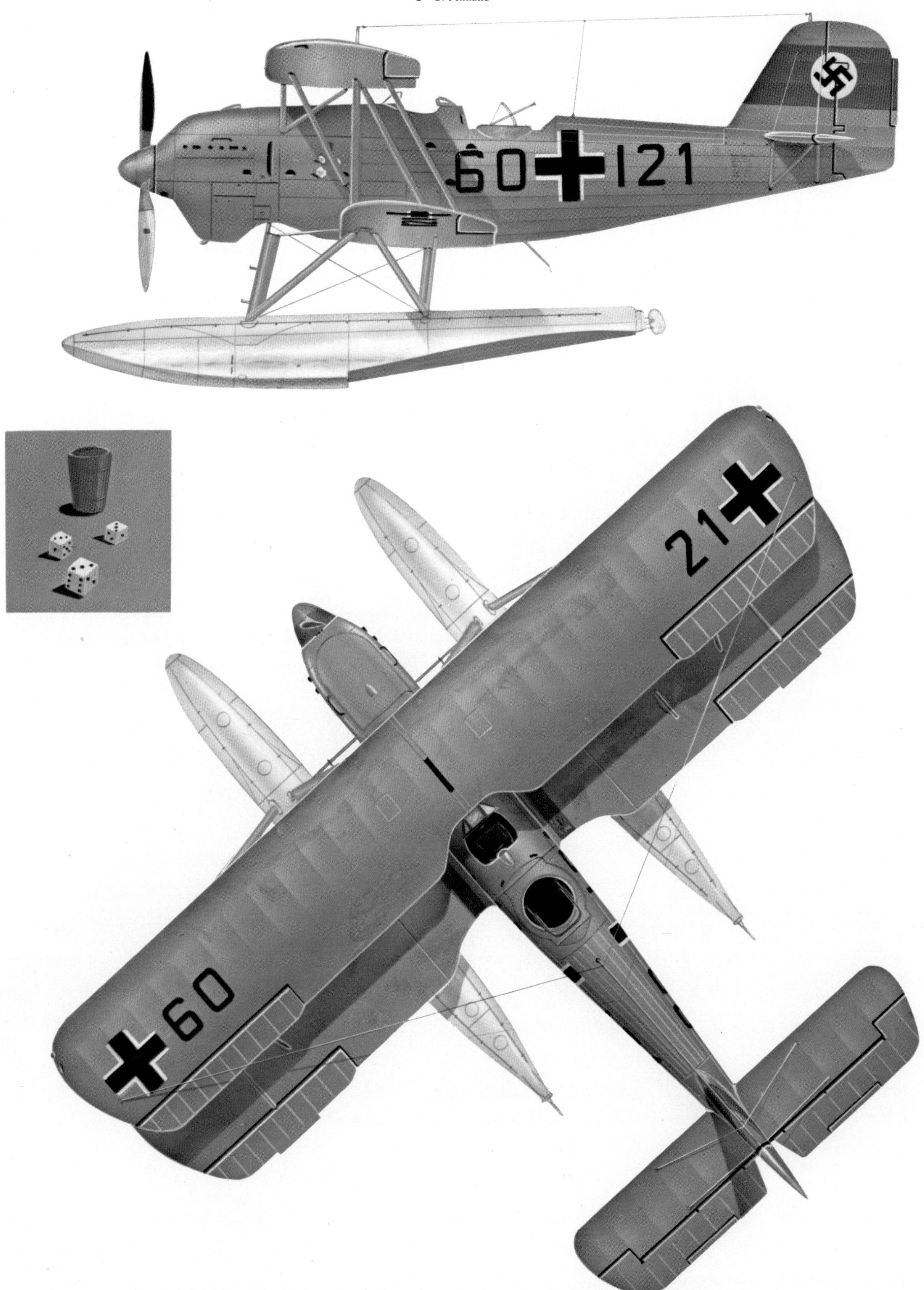

Finished in overall *RLM* 02, this He 60C of 1 /Kü Fl Gr 206 provides an interesting comparison with the one illustrated on Page 50. Note the different style of the actual code application. Also, on the aircraft above, the code was aligned parallel to the ground when the aircraft was at rest. Due to the complexity of the floats and struts, the last three characters of the code, I21, were grouped beneath the lower port wingtip. The floats were finished in overall silver colored aluminium paint, a practice later limited to surfaces below the waterline. The *Staffel* emblem appeared on both sides of the fuselage.

Typical of the early production home-based Bf 109s, this aircraft of II Kü J Gr/136 was finished in a color scheme of 70/71/65. Its identifying code number was applied in very large white numerals without black edging, a feature of fighter markings when the new very dark color scheme was introduced.

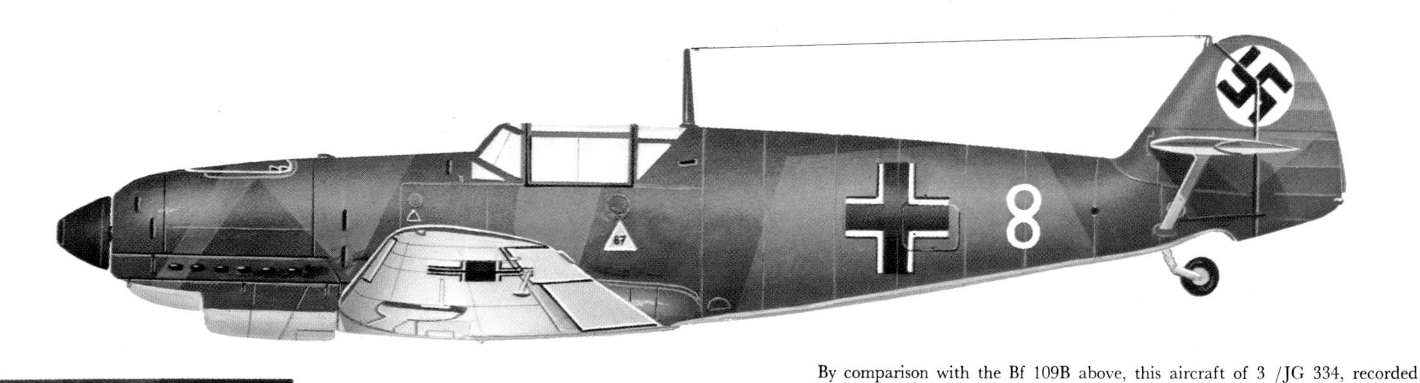

By comparison with the Bf 109B above, this aircraft of 3 /JG 334, recorded nearly a year later in late 1938, carried its identifying white number in the correct proportions of 6/10 the height of the fuselage *Balkenkreuz,* but technically in the wrong location.

Bf 109E, W Nr 3272, of IV /JG 132 just prior to the outbreak of hostilities. Camouflaged in 70/71/65, it carried the relatively rare IV *Gruppe* marking in combination with the earlier style of very large white identifying number without black edging. The boot emblem was applied to both sides of the fuselage and was retained as the *Gruppe* emblem when the unit was redesignated I /JG 77.

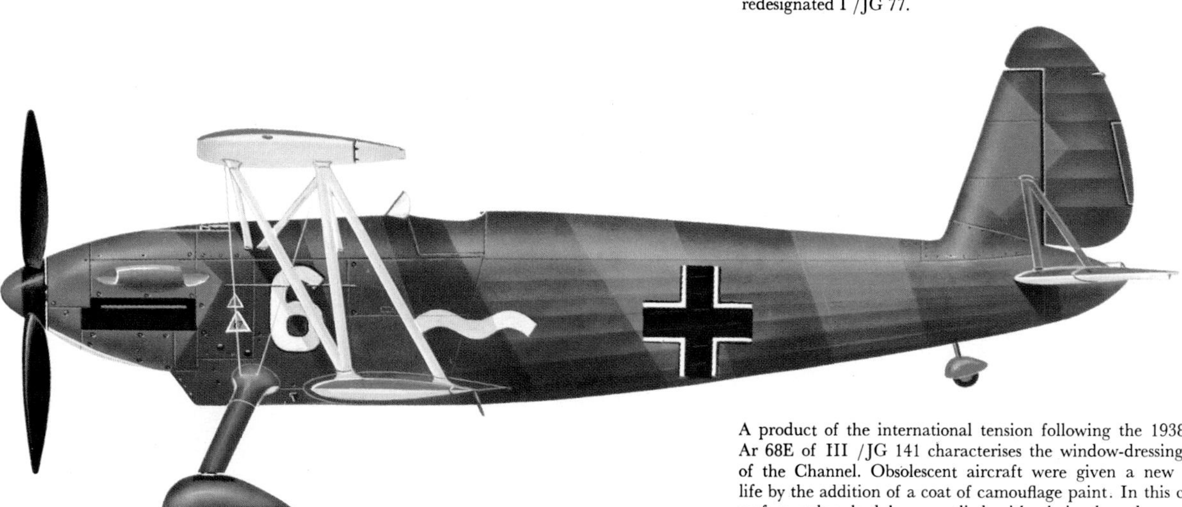

A product of the international tension following the 1938 Munich crisis, this Ar 68E of III /JG 141 characterises the window-dressing done on both sides of the Channel. Obsolescent aircraft were given a new lease of operational life by the addition of a coat of camouflage paint. In this case the 70/71 upper surface colors had been applied with obviously only vague reference to the standard pattern for single-engined fighter aircraft. In the process, the *Hakenkreuz* had been completely deleted. The III *Gruppe* marking was also the very early shallow type.

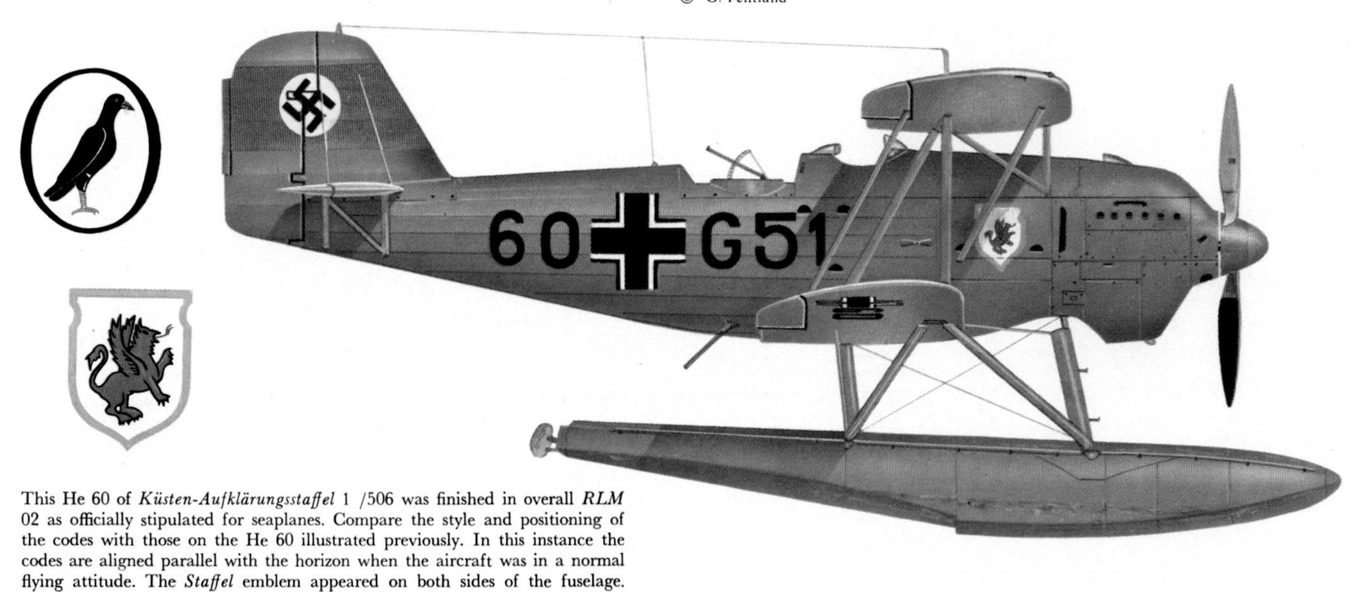

This He 60 of *Küsten-Aufklärungsstaffel* 1 /506 was finished in overall *RLM* 02 as officially stipulated for seaplanes. Compare the style and positioning of the codes with those on the He 60 illustrated previously. In this instance the codes are aligned parallel with the horizon when the aircraft was in a normal flying attitude. The *Staffel* emblem appeared on both sides of the fuselage.

A Do 18 of 2 /Kü Fl Gr 406, just prior to the Munich crisis of 1938, in standard finish for flying boats. The color scheme was *RLM* 02 for all surfaces other than the undersides of the hull and sponsons which were painted with one coat of 01 and two coats of clear lacquer 00. The division between the two surfaces was marked with the regulation 28-colored 3-mm line positioned 10-cm above the high water line. The thin white lines just below and forward of the cockpit side window were sighting lines for use with the hand-held reconnaissance camera. The *Staffel* emblem appeared on both sides of the engine cowling.

An interesting example of bomber unit markings prior to the outbreak of war. This Do 17 of I /KG 153 carried its full code in black—contrary to prescribed standards. The spinners were painted in the red *Staffel* color and a white band around the nose of the fuselage denoted I *Gruppe*. The aircraft's individual letter was repeated in black on the nose of the aircraft. In common with other aircraft of this particular period, the red background and white disc to the *Hakenkreuz* had been overpainted, in this instance with 61. The *Gruppe* emblem appeared on both sides of the fuselage but was slightly further aft on the starboard side due to the extensive glazing. The camouflage scheme was standard 61/62/63/65.

Recorded during the closing days of Summer 1939, this Bf 110 of I /ZG 76 provides a good example of conformity to all the then current *RLM* directives. The color scheme was 70/71/65 and the national markings were of correct style, size, proportion and positioning. Unlike many other front-line aircraft of this period, the aircraft's individual letter had been correctly applied in the *Staffel* color.

A brace of Ju 86s photographed during the manoeuvres held in 1937. The nearer aircraft, 42 + C26, was an A-1, the other a D-1, both being finished in the standard camouflage of 61/62/63/65. The broad band around the nose of each aircraft appears to have been yellow. The weathered nose band of the nearer aircraft shows the nature of the glyptal resin-based temporary distemper which was meant to be used only for the duration of the exercises. Note that the spinner tips were also painted in the same temporary yellow color.

Another Ju 86D, also camouflaged in 61/62/63/65, but with the upper surface colors taken right down to the lower extremities of the fuselage sides. The Junkers-Hamilton propeller blades were left in natural metal and bore the oval-shaped manufacturer's trade mark.

and 7230. These paints were used on all exterior wooden parts with the exception of seaplane floats and flying boat hulls. (After the outbreak of war, 7135 was also listed for use in conjunction with any of the standard camouflage *Farbtöne*. 7230 was also eventually replaced by 7236.)

Flieglackkette Nr 31 — This group consisted of *Flieglack* 7131, 7132, 7135 and 7230. *Farbton* 01 was used in conjunction with 7135 and 00 with 7131 and 7230. Uses and restrictions applicable to this group were the same as those for *Flieglackkette Nr 30.* (After the outbreak of war this group was gradually phased out).

Flieglackkette Nr 32 — This group consisted of *Flieglack* 7145 and 7162 plus protective finishes 7180 and 7181. *Farbtöne* 00 and 01 were used in conjunction with 7162, 02 with 7145, and 22 with 7180. The uses of these paints and protective finishes were restricted to seaplane floats and flying boat hulls as follows: Internally all surfaces were finished with 7180.22. Externally, all surfaces above the waterline were first treated with 7181 and then given two coats of 7145.02; all surfaces below the waterline were given one coat of 7162.01 followed by two coats of 7162.00. (With the outbreak of war this group was rapidly phased out).

Flieglackkette Nr 33 — This group consisted of *Flieglack* 7115, 7138, 7139 plus 7213 and 7215. *Farbton* 02 was used in conjunction with 7115 and any of the other standard camouflage *Farbtöne*. All paints in this group were non-inflammable as their use was specifically for fabric coverings on shipborne aircraft. Paints in this group were not interchangeable with those from *Flieglackkette Nr 30.* (This group was introduced after the outbreak of war.)

The *Farbtöne* listed in the above information were issued in the form of a standard color chip chart. The earliest known example of this chart is dated 1937 and is the only one to list the *Farbtöne* 61, 62 and 63. By the time the second issue was released in 1938, these three colors had been officially deleted for reasons which will be explained. (A full listing of all the colors contained in these two issues of L Dv 521 are to be found later in this chapter.)

In 1936 a new aircraft type designed specifically to meet the requirements for dive-bombing, appeared in the form of the Henschel Hs 123. The first five production aircraft were shipped to Spain for field evaluation in December of that year, and were among the first aircraft to carry the new segmented camouflage scheme as mentioned in Chapter 2. The actual camouflage scheme utilised the existing pale greenish-grey (identified by the code 63), to which was added the dark brown (61) and green (62), all undersurfaces being painted blue-grey (65). The division between upper and lower colors was quite sharply delineated.

The introduction of the new colors corresponded with the issue of official camouflage patterns which were then incorporated into the various aircraft technical maintenance manuals. Similar to British policy for its RAF aircraft, alternative variations were made possible by the use of what was termed a mirror image of the standard pattern. Unlike the normal British system however, a further degree of variation was achieved by the transposition of two of the three colors used for the upper surfaces. Thus six distinct schemes were available.

The actual camouflage patterns were composed of straight lines set out on a basic grid pattern of squares. The total area set out was of sufficient proportion to cover entirely a superimposed plan view of any standard *Luftwaffe* aircraft type. The usual practice was to project the grid so that it just enclosed the full span of the particular aircraft. Variations in the position, shape and area of the wing and tailplane horizontal surfaces produced the seemingly marked differences in the camouflage pattern seen on aircraft of this period.

The Hs 123 was the subject of several camouflage variations. While it was common practice to lay out the camouflage scheme with the longitudinal centre line of the aircraft type aligned with the centre line of the diagram, the Hs 123 pattern was offset to one side by one grid square. In addition, the actual pattern also incorporated several modifications and minor additions as illustrated. Being a biplane, the upper surfaces of the bottom wings were camouflaged, the pattern being identical with the corresponding area on the upper surfaces of the top wings.

Fighter color schemes remained unchanged at this time and a full painting schedule for the He 51 is reproduced on the following pages as an example of the extent and complexity of the paint applications of this period. One point that is of particular significance on Page 1 of the painting schedule is the use of the identification *grüngrau* 63. The official *RLM* listing identified 63 as *hellgrau,* but the first description was, in fact, more appropriate. Official designations were somewhat confusing at times and the color descriptions given them should be treated with due caution.

The absence of a fighter camouflage scheme however, was soon to be remedied. The Messerschmitt Bf 109 had already made its service début in Spain and supplies to this theatre of operations slowed down the equipping of the home-based squadrons. Thus by mid-1937 only I and II/JG 132 and I/JG 234 were equipped with the type. Their introduction into the home-based fighter strength coincided with the introduction of a camouflage scheme for fighter aircraft. Like the bomber scheme, a straight-edged splinter pattern was used but only two colors were used; black-green (70) and dark green (71). The undersurfaces used blue-grey (65) as on the bombers. The two greens were of an extremely dark shade with a correspondingly low contrast value. (See Page 110). The division line between upper and lower surface colors was kept as low as possible on the fuselage sides. All the standard nationality and military markings were retained, but stencilled instructions now appeared in wine-red (28) on the green areas and in black (22)

A line-up of Hs 123s of 3 /St G 165 *Immelman* photographed in 1937. Although standard in all other respects as regards the camouflage and markings of the period, the type was unique in the application of the aircraft's individual letter in white. Also, unlike the early model Ju 87s, the upper surface colors were not taken right down the fuselage sides and terminated instead in a relatively high demarcation line.

W Nr 968 of 3 /St G 165 shows the intricate upper surface camouflage pattern and the repetition of the aircraft's identification letter on the top wing upper surface centre section. Unlike its contemporaries, the Hs 123 did not carry the full code sequence on its top wing upper surfaces.

Farbanstrich für Flugzeugmuster He 51 C,D

(Dichtungs- und Isolationsmaterial ist mit anzugeben)

Blatt: 1
Diese Anstrichliste enthält **6 Blatt.**

Flugzeug Teilbezeichnung	Baustoff	Anstrich Zahl	Streichen-Spritzen	Trocknung Ofen Std.	Trocknung Luft Std.	Musterbezeichnung des Anstrichmittels (nur nach Angabe auf der Kanne)	Hersteller	Bemerkung
Rumpfwerk								
Rohrgerüst	Stahlrohr	1	spritzen		6	Avionorm Ölgrund hellgrau Nr.5759	Lüdicke & Co.	
		1	spritzen		4	" Nitro-Decklack silber Nr.1708	" "	
		1	spritzen		4	" Nitro-Lasurlack silber V 192	" "	
			ausgiessen			Sinoxydöl gemischt mit 7% Siccativ-Extrakt	Chemie-Farben	Nicht bei reinen Landmaschinen
Umwicklung	Leinenband	1	streichen			50% Parafin, 30% Stearin, 20% Bienenwachs		Anstrich unter Umwicklung.
Formgebungsgerüst	Kiefer	2	spritzen je		6	Bakelite-Firnis R	Bakelite G.m.b.H.	
Rumpfhauben	Aluminium, Hydronalium	1	spritzen		6	Avionorm Ölgrund hellgrau Nr.5759	Lüdicke & Co.	
		1	spritzen		4	" Nitro-Decklack silber Nr.1708	" "	
		2	spritzen je		4	" Nitro-Decklack grau Nr.7007	" "	
	Elektron	1	spritzen		6	Avionorm Ölgrund grau Nr.5706	Lüdicke & Co.	
		1	spritzen		4	" Nitro-Decklack silber Nr.1708	" "	
		2	spritzen je		4	" Nitro-Decklack grau Nr.7007	" "	
Bespannstoff	Weber-Leinen 1403 FL II (oder Kisker 11 FO IV)	3	streichen je leicht schleifen		2 =	Cellesta-N-Flugzeuggrundl.rot Nr.1603B Schleifpapier	Atlas Ago L.M.	
		2	spritzen je verteilen mit		4	Cell.Flugzeug.Übz.L.Nr.2000 grüngr.63 Ausgleichsfl.Nr. 1611	" " " / " " "	Der 4. 5.u.6.Anstrich ist m. 10% Ausgleichsfl.Nr.1611 zu verdünnen
		1	spritzen		4	Cell.Flugzeug.Übz.L.Nr.2000 grüngrau 63	" " "	
		1	spritzen		6	mit einem Gemisch aus		
						5 R.T.Cell.Flugzeug.Übz.L.Nr.2000 grüngrau	" " "	
						2 R.T. " " " " Nr.1606 farblos	" " "	
						3 R.T. " Ausgleichsfl.Nr. 1611	" " "	
Fahrwerk								
Sämtliche Streben	Stahlrohr	1	spritzen		6-7	Avionorm Ölgrund hellgrau Nr.5759	Lüdicke & Co.	
		1	spritzen		4	" Nitro-Decklack silber Nr.1708	" "	
		2	spritzen je		4	" Nitro-Decklack grau Nr.7007	" "	
innen			ausgiessen			Sinoxydöl gemischt mit 7% Siccativ-Extrakt	Chemie-Farben	Bei Land- und Seemaschinen.
Verkleidung	Elektron	1	spritzen		6	Avionorm Ölgrund grau Nr.5706	Lüdicke & Co.	
		1	spritzen		4	" Nitro-Decklack silber Nr.1708	" "	
		2	spritzen je		4	" Nitro-Decklack grau Nr.7007	" "	

Farbanstrich für Flugzeugmuster He 51 C,D

(Dichtungs- und Isolationsmaterial ist mit anzugeben)

Blatt : 2
Diese Anstrichliste enthält
6 Blatt.

Flugzeug Teilbezeichnung	Baustoff	Anstrich Zahl	Streichen- Spritzen	Trocknung Ofen Std.	Trocknung Luft Std.	Musterbezeichnung des Anstrichmittels (nur nach Angabe auf der Kanne)	Hersteller	Bemerkung
Schwimmerwerk								
Schwimmer innen Spanten und Beplankung	Hydronalium	2	spritzen je		3	Avionorm Bitumen-Lack silber Nr.120	Lüdicke & Co.	
Dichtungsmittel	Schadebinde						Chemierprodukte G.m.b.H.	
Schwimmer aussen	Hydronalium							
Grundierung		1	spritzen		12	Avionorm Ölgrund grau extra fett Nr.230 über die ganze Aussenhaut.	Lüdicke & Co.	
Deckanstrich								
unter Wasser		2	spritzen		3-4	Avionorm Bitumen-Lack silber Nr.120	Lüdicke & Co.	
über Wasser		1	spritzen		4	" Nitro-Decklack silber Nr.1708	" "	
		1	spritzen		4	" Nitro-Lasurlack silber Nr.V 192	" "	
Beschläge	Stahlblech	1	spritzen		12	Avionorm Ölgrund grau extra fett Nr.230	Lüdicke & Co.	
Knotenstücke	Stahlguss	1	spritzen		4	" Nitro-Decklack silber Nr.1708	" "	
		1	spritzen		4	" Nitro-Lasurlack silber Nr.V 192	" "	
Schwimmerstreben	Stahlrohr	1	spritzen		6	Avionorm Ölgrund hellgrau Nr.5759	Lüdicke & Co.	
		1	spritzen		4	" Nitro-Decklack silber Nr.1708	" "	
		2	spritzen		4	" Nitro-Lasurlack grau 7007	" "	
Strebenverkleidungen	Hydronalium							
Kielleisten vor der Stufe	Hydronalium	1	spritzen		12	Avionorm Ölgrund grau extra fett Nr.230	Lüdicke & Co.	
		2	spritzen je		3	" Bitumenlack silber Nr.120	" "	
hinter d.Stufe		2	spritzen je		3	" Bitumenlack silber Nr.120	" "	
Schwimmerstreben innen			ausgiessen			Sinoxydöl gemischt mit 7% Siccativ-Extrakt	Chemiefarben	
Tankräume		1	spritzen		12	Avionorm Ölgrund grau Nr.5706	Lüdicke & Co.	
		1	ziehen			" Nitro-Spachtel weiss Nr.7587	" "	nur auf den Dichtnähten.
		1	spritzen		4	" Nitro-Decklack silber Nr.1708	" "	
		1	spritzen		4	" Nitro-Lasurlack silber Nr.V 192	" "	

Farbanstrich für Flugzeugmuster He 51 C,D

(Dichtungs- und Isolationsmaterial ist mit anzugeben)

Blatt : 3
Diese Anstrichliste enthält
6 Blatt.

Flugzeug Teilbezeichnung	Baustoff	Anstrich Zahl	Streichen-Spritzen	Trocknung Ofen Std.	Trocknung Luft Std.	Musterbezeichnung des Anstrichmittels (nur nach Angabe auf der Kanne)	Hersteller	Bemerkung
L e i t w e r k								
Höhenflossen, Höhenruder								
Höhenhilfsruder								
Seitenflosse, Seitenruder								
Querruder								
Gerüst	Hydronalium							
Tragende Bepl.								
innen		1	spritzen		6	Avionorm Ölgrund hellgrau Nr.5759	Lüdicke & Co.	
aussen (soweit unter		1	spritzen		4	" Nitro-Decklack silber Nr.1708	" "	
Bespannung)		1	spritzen		4	" Nitro-Lasurlack silber V 192	" "	
aussen		1	spritzen		6	Avionorm Ölgrund hellgrau Nr.5759	Lüdicke & Co.	soweit die Teile von außen
(soweit ohne Bespannung)		1	spritzen		4	" Nitro-Decklack silber Nr.1708	" "	sichtbar sind, wird m.Lasur-
		2	spritzen	je	4	" Nitro-Decklack grau Nr.7007	" "	lack v.Atlas-Gao überspritzt.
Tragende Beplankung, soweit später mit Stoffbespannung in Berührung kommt.		1	streichen			Avionorm Zwischenlack rot Nr.6663	Lüdicke & Co.	
Teilbeplankung	Weber-Leinen 1403 FL II	3	streichen leicht schleifen	je	2	Cellesta N-Flugzeuggrundl.rot Nr.1603 B mit Schleifpapier	Atlas-Ago L-M.	
	(oder Kisker	2	spritzen	je	4	Cell.Flugzeug Übz.L.Nr.2000 grüngr.63	" " "	der 4.,5.u.6.Anstrich ist
	11 FO IV)		verteilen mit			Ausgleichsflüssigkeit Nr.1611	" " "	mit 10% Ausgleichsflußigkeit
		1	spritzen		4	Cell.Flugzeug Übz.L.Nr.2000 grüngr.63	" " "	Nr.1611 zu verdünnen.
		1	spritzen		6	mit einem Gemisch aus		
						5 R.T.Cell.Flugzeug Übs.L.Nr.2000 grüngrau 63	" " "	
						2 R.T. " " " " " 1606 farblos	" " "	
						3 R.T. " Ausgleichsfl.Nr.1611	" " "	

Farbanstrich für Flugzeugmuster He 51 C,D

(Dichtungs- und Isolationsmaterial ist mit anzugeben)

Blatt: 4
Diese Anstrichliste enthält
6 Blatt.

Flugzeug Teilbezeichnung	Baustoff	Anstrich Zahl	Streichen- Spritzen	Trocknung Ofen Std.	Trocknung Luft Std.	Musterbezeichnung des Anstrichmittels (nur nach Angabe auf der Kanne)	Hersteller	Bemerkung
Tragwerk								
Holme	Kiefer und	2	streichen je		12	Yachtlack superfein	Herbig-Haarhaus	
Rippen	Sperrholz							
Holme innen			ausgiessen			Yachtlack superfein	Herbig-Haarhaus	
Innenverspannung	Stahldraht		streichen		12	Yachtlack superfein mit Zusatz blauer Farbe	Herbig-Haarhaus	
Beschläge	Stahlblech	1	spritzen		6	Avionorm Ölgrund hellgrau Nr.5759	Lüdicke & Co.	
		1	spritzen		4	" Nitro-Decklack silber Nr.1708	" "	
		1	spritzen		4	" Nitro-Lasurlack silber V 192	" "	
Beplankung	Weber-Leinen 1403 FL II	3	streichen je leicht schleifen		2 mit	Cellesta-N-Flugzeuggr.L.rot N.1603 B Schleifpapier	Atlas-Ago L-M.	der 4.,5.u.6.Anstrich ist m. 10% Ausgleichsfl.Nr.1611 zu
	(oder Kisker	2	spritzen je		4	Cell.Flugzeug Übz.L.Nr.2000 grüngr.63	" " "	verdünnen.
	11 FO IV)		verteilen mit			Ausgleichsflüssigkeit Nr.1611		
		1	spritzen		4	Cell.Flugzeug Übz.L.Nr.2000 grüngr.63	" " "	
		1	spritzen		6	mit einem Gemisch aus		
						5 R.T.Cell.Flugzeug Übz.L.Nr.2000 grüngr.63	" " "	
						2 R.T. " " " " " 1606 farblos	" " "	
						3 R.T. " Ausgleichsfl.Nr.1611	" " "	
		1	streichen		3	Cell.Holzgrdg.farbl.Nr.2254	Atlas-Ago L-M.	
		2	spritzen je		5	Cell.Holzfül.grau Nr.2070	" " "	der Füller ist mit 25% Ausgl. fl.Nr.1611 zu verdünnen.
			ausgleichen		mit	Cell.Ausgleichsfl.Nr.1611 T.Z. 2h	" " "	
		2	spritzen je		4	Cell.-N-Flugzeug Übz.L.Nr.2000 grüngr.63	" " "	
Teilbeplankung	Sperrholz	1	spritzen		5	Cell.-N-Flugzeug Übz.L.Nr.2000 grüngr.63 50%	" " "	
						Cell.-N-Flugzeug Übz.L.Nr.1606 farbl. 20%	" " "	
						Cell.Ausgleichsfl.Nr.1611 30%	" " "	
Soweit später mit Stoffbespannung in Berührung kommt.		1	streichen			Avionorm Zwischenlack rot Nr.6663	Lüdicke & Co.	
Trittbeplankung	Sperrholz	2	spritzen je		6	Bakelite Firnis F.	Bakelite G.m.b.H. Erkner	
		2	spritzen je		4	Avionorm Nitro-Decklack hellgrau Nr.1791 säurefest	Lüdicke & Co.	

Farbanstrich für Flugzeugmuster He 51 C,D

(Dichtungs- und Isolationsmaterial ist mit anzugeben)

Blatt : 5
Diese Anstrichliste enthält
6 Blatt.

Flugzeug Teilbezeichnung	Baustoff	Anstrich Zahl	Streichen - Spritzen	Trocknung Ofen Std.	Trocknung Luft Std.	Musterbezeichnung des Anstrichmittels (nur nach Angabe auf der Kanne)	Hersteller	Bemerkung
Stiele aussen	Stahlrohr	1	spritzen		6	Avionorm Ölgrund hellgrau Nr.5759	Lüdicke & Co.	
		1	spritzen		4	" Nitro-Decklack silber Nr.1708	" "	
Spannbock aussen		2	spritzen	je	4	" Nitro-Decklack grau Nr.7007	" "	
Stiele innen			ausgiessen			mit Sinoxydöl gemischt mit 7% Siccativ-Extrakt	Chemie-Farben	Bei Land-u.Wassermaschinen.
Spannbock - Auskreuzung	Profildraht		streichen		12	Yachtlack superfein mit Zusatz blauer Farbe	Herbig- Haarhaus	
Klappen in der Beplankg.	Duraluminium	1	spritzen		6	Avionorm Ölgrund grau Nr.5759	Lüdicke & Co.	
		1	spritzen		4	" Nitro-Decklack silber Nr.1708	" "	
		2	spritzen	je	4	" Nitro-Decklack grau Nr.7007	" "	
Steuerwerk								
Stoßstangen, Beschläge, Hebel, Lagerböcke, Rollen	Duraluminium, Stahlblech, Siluminguss	1	spritzen		6	Avionorm Ölgrund hellgrau Nr.5759	Lüdicke & Co.	
		1	spritzen		4	" Nitro-Decklack silber Nr.1708	" "	
		1	spritzen		4	" Nitro-Decklack (Lasur) silber Nr.V 192	" "	
	Elektron	1	spritzen		6	Avionorm Ölgrund grau Nr.5706	Lüdicke & Co.	
		1	spritzen		4	" Nitro-Decklack silber Nr.1708	" "	
		1	spritzen		4	" Nitro-Lasurlack silber V 192	" "	
Züge	Stahlrunddraht					Kadmier:		
Triebwerk								
Motorräume								
Gerüst	Stahlrohr	1	spritzen		6	Avionorm Ölgrund hellgrau Nr.5759	Lüdicke & Co.	
Verkleidung innen	Aluminium	1	spritzen		4	" Nitro-Decklack silber Nr.1708	" "	
Bedienungsgestänge	Dural.	1	spritzen		4	" Nitro-Lasurlack silber V 192	" "	
Behälter								
Für Kraftstoff								
Für Schmierstoff								
Für Kühlstoff								
Brandschott Führerseite								

Farbanstrich für Flugzeugmuster He 51 C,D

(Dichtungs- und Isolationsmaterial ist mit anzugeben)

Blatt: 6
Diese Anstrichliste enthält 6 Blatt.

Flugzeug Teilbezeichnung	Baustoff	Anstrich Zahl	Streichen- Spritzen	Trocknung Ofen Std.	Trocknung Luft Std.	Musterbezeichnung des Anstrichmittels (nur nach Angabe auf der Kanne)	Hersteller	Bemerkung
Brandschott Motorseite						ohne Anstrich		
Motorverkleidung aussen		1	spritzen		6	Avionorm Ölgrund hellgrau Nr.5759	Lüdicke & Co.	
		1	spritzen		4	" Nitro-Decklack silber Nr.1708	" "	
		2	spritzen	je	4	" Nitro-Decklack grau Nr.7007	" "	
Rohrleitungen								
Für Kraftstoff	Aviotub-Schl.		streichen		3	Kennzeichnung gelb Nr.2099	Lüdicke & Co.	
Für Schmierstoff	" "		streichen		3	Kennzeichnung braun Nr.2181	" "	
Für Kühlstoff	" "		streichen		3	Avionorm-Decklack grün Nr.2183	" "	
Für Feuerlöschanlage	" "		streichen		3	" Decklack rot Nr.7150	" "	
Für Luft	" "		streichen		3	" Decklack blau Nr.6686	" "	
Instrumentenbrett	Elektron	2	spritzen	je	2	Avionorm Decklack Nr.702/3850 grau-matt	Lüdicke & Cc.	
Akkulagerung		1	spritzen		6	Avionorm Ölgrund grau Nr.5759	Lüdicke & Co.	
		1	spritzen		3	" Decklack hellgrau Nr.1791 säurefest	" "	
Beschriftung								
D - Nummern		1	streichen		12	Elastadur schwarz	Obermann - Berlin	
Hoheitszeichen		2	streichen	je	3	Avionorm rot Nr.7150	Lüdicke & Co.	
		2	streichen	je	3	" schwarz Nr.1730	" "	
		1	streichen		3	" weiss Grund V 18	" "	
		1	streichen		3	" weiss Nr.1859	" "	

Used as an engine test bed, the Do 17 V10 eventually had its civil registration D-AKUZ deleted and a 61/62/63/65 camouflage scheme applied in place of its original overall 63 color. The division between upper and lower surface colors was standard for Do 17s.

This He 111B demonstrates the apparent variation in upper surface camouflage patterns. The variation was in fact not due to a different camouflage pattern so much as the relative position of the wings as explained in the main text. The extreme outboard positioning of the *Balkenkreuze* is clearly evident.

on the areas painted 65.

Many references have been made to Bf 109s of this period being painted in an overall scheme of either of the two greens mentioned above, but there is little evidence to substantiate this theory. Most photographic 'examples' of this apparent overall green scheme are almost always a result of the extremely low contrast values of the two greens and the limitations of certain photographic emulsions in use at the time. While it is an accepted fact that during the vagaries of war, field units of the *Luftwaffe* were to exercise a certain justifiable laxity towards the application of some camouflage directives, such was the standard of control during those early pre-war years that failure to comply with the prescribed directives was simply not tolerated. For those who doubt the strictness of the control exercised in this period, they should consult any of the surviving directives for the painting of aircraft, which were liberally laced with reminders *ad nauseam* to obey instructions to the letter. The tensions of the immediate pre-war period were to produce some temporary anomolies regarding markings, but order still reigned in the period under discussion.

Photographs of very early production Bf 109s show the 70/71/65 camouflage scheme in use, a combination that was to last in this form until early 1940. The general public had its first close look at the Bf 109 during the 4th International Flying Meeting which commenced at Zürich's Dübendorf airfield on 23rd July 1937. Four of the five Bf 109s were pre-production prototypes and bore either the old overall 63 finish with black civil registration or a more stylish racing color and trim. The fifth machine however, was a production B-2 model and was fully camouflaged in the 70/71/65 scheme. Despite an extensive search of surviving documentation, nothing has been found that specifies the use of a single upper surface color scheme in either of the two greens 70 or 71. There would appear to have been little justification or sense in repealing the existing camouflage directive, issuing a new one, only to re-instate the original one in a period of just over a year.

One further point requires elucidation. The camouflage colors under discussion have been consistently referred to in the past as having a matt finish. Except for certain special finishes, normal camouflage paints had a *flat* finish which is something distinctly different, and these did produce a small amount of reflection under certain conditions. Thus German aircraft varied considerably from their British/American counterparts. Zinc oxide primer was used for the German paints in place of the more usual zinc chromate for metal parts, this choice probably being dictated originally by the lack of necessary home-based mineral resources. The zinc oxide variety also had the advantage of being less costly, an important factor when one considers the large quantities in question. Fabric-covered surfaces were doped with a clear lacquer and then enamel of a non-oxidizing type was applied directly over both the lacquer and the zinc oxide coated metal surfaces. The application of enamel over the lacquered surfaces protected them from deterioration due to ultra-violet light.

The non-oxidizing enamel had the advantage of being easy to remove with the aid of toluol or a similar stripper and without damaging the lacquered surfaces underneath. British oxidizing alkyd-resin enamels could not be removed cleanly, leaving a film of enamel which was both hard to remove and impossible to repaint successfully. The only disadvantage of the German

A factory-fresh Bf 109B-1 seen in Summer 1937, in the then new fighter camouflage colors of 70/71/65. The low contrast value between the two greens 70 and 71 has produced many of the photographic 'examples' of what has often been claimed as an overall scheme of 70 or 71.

A pair of Bf 109Bs from II/JG 132 at Jüterborg-Damm in the summer of 1937. They display the new 70/71/65 fighter scheme with large white identity numbers. When dark camouflage was introduced the old practice of outlining the aircraft's identity number thinly in black ceased, since there was now sufficient contrast with the background. The actual size of the numbers in proportion to the *Balkenkreuz* was also altered to make them both the same height. This practice appears to have been limited in its application to a few units. The red script *R* on a white shield was the unit's emblem and appeared on both sides of the fuselage. *Personal emblems when carried on Luftwaffe aircraft were usually confined to the port side, the traditional side from which either a horse or an aircraft was mounted!*

Three He 111B-2s respendent in full military markings and 61/62/63/65 camouflage. The lead aircraft was coded 71 + H26, the rearmost machine being K and the other G.

Two shots of a pre-production Ju 87B-0 bearing the civil registration D-ILEX and camouflaged in 61/62/63/65. The small white rectangle just forward of the letter I covered an emergency first aid kit.

enamels was the comparatively low pigmentation density which necessitated the use of several coats and thus proved heavier per unit area than the comparable British product.

1938 was to prove a fateful year politically and one in which the uncertainty of the times was to be reflected by the ferment of change in *Luftwaffe* camouflage and markings. Pressure for the reunification of the predominantly German-populated Sudaten area of Czechoslovakia with the greater *Reich* territory resulted in the August crisis. While the world metaphorically held its breath, the city of Munich became the venue for rapid negotiations to forestall the threat of war. Tension eased temporarily and the British Prime Minister returned to England bearing an agreement that was to buy both sides one year of grace. During that year a rapid expansion of military forces was undertaken on both sides of the Channel.

This crisis was followed by a rapid change in both camouflage and markings. The newer scheme of 70/71/65 in use for fighters was deemed more suitable to a continental conflict than the older scheme of 61/62/63/65, and future deliveries from the aircraft factories were to be seen bearing the later scheme. The change however, involved more than just the colors, and a new splinter pattern as adopted for the Bf 109 was instituted. Like the earlier three-color pattern the two-color pattern was laid out in a basic form covering an area in excess of that required. Again this pattern could be used in standard form or as a mirror image, but transposition of the two colors 70 and 71 does not appear to have been instituted at this time, although it was to be seen in use around 1940.

Preparation of the official diagrams for each aircraft type was done by projection of the grid and pattern over the basic shape. Like the earlier pattern, the actual shapes and areas, particularly those of the wings, produced the seeming variations between the types. Unlike the earlier types of diagrams however, full side elevations were included to show how the scheme should be applied to the sides of the aircraft. Sample diagrams are shown on Page 76.

The decision to implement the change took less time than the actual physical aspects involved in repainting, and initially some seemingly anomolous sights were seen. Accompanying the general adoption of the 70/71/65 camouflage scheme was an order to delete the red band and white disc background to the *Hakenkreuz* since these now compromised the general camouflage.

Coastal reconnaissance aircraft were also allocated a camouflage pattern based on the standard one adopted for land-based aircraft with the colors 72 and 73 substituted for 70 and 71 respectively. Some of the He 114s however, appear to have at least initially retained their general scheme of overall 02, the only concession to the revised markings being the deletion of all but the actual *Hakenkreuz* and the adoption of white for the aircraft's individual code letter.

With the introduction of the 70/71/65 scheme for bomber and reconnaissance aircraft, visual indentification by codes now became virtually impossible, except at very close quarters. This was due to the very low contrast between the camouflage colors and the black of the codes. To alleviate this problem the third and fifth characters of the code were repainted in white. Thus the aircraft's identity and its *Staffel* were readily discernable, a prerequisite for formation tactics.

The *RLM* had already issued extensive technical notes concerning paints and their application, and 1938 was to see a further extension of this information. The existing official colors were intended for multi-purpose application and are listed below by their *Farbton* code number, official designation where known, followed by a brief note on their main uses. The final code number in brackets refers to the 1970 Munsell Matte Book of Color. This reference has been chosen because of its international application and its precise method of color identification. The actual matching was done against original editions of L Dv 521. The reader is cautioned however about the Munsell references quoted for the colors 61, 62 and 63, which were matched from a different source. While this source is recognised as being reputable, the precise shades have yet to be verified against an official *RLM* document. The Munsell reference should therefore be treated as being approximate.

00 — *wasserhell* — (colorless) — used as a protective finish.

01 — *silber* — (silver) — used for interior and exterior finishes.

02 — *RLM-grau* — (*RLM*-grey) — used as an external finish for floatplanes and flying boats. Also used in some instances as a primer coat. (5GY 5/1.)

03 — *silber* — (silver) — used for external finishes.

04 — *gelb* — (yellow) — used for identification markings and fuel markings. (2.5Y 8/12.)

05 — — (creamy yellow) — used as a gloss external finish for sailplanes and gliders. (5Y 8/6.)

21 — *weiss* — (white) — used for identification markings. (N9.5 90%R.)

22 — *schwarz* — (black) — used for identification markings. (N2.75 5.5%R.)

23 — *rot* — (red) — used for identification markings. (7.5R 4/12.)

24 — *dunkelblau* — (dark blue) — used for internal systems identification markings. (5PB 2.5/6.)

25 — *hellgrün* — (light green) — used for identifica-markings. (5G 5/6.)

26 — *braun* — (brown) — used for internal systems identification and oil markings. (2.5YR 4/6.)

27 — *gelb* — (yellow) — used for identification markings. (5Y 7/10.)

28 — *weinrot* — (wine-red) — used for walkway markings. (10RP 2.5/4.)

61 — *dunkelbraun* — (dark brown) — used as camou-

A He 111E-0 camouflaged in 61/62/63/65. The propeller blades were painted pale grey, those on the starboard engine bearing the Junkers emblem while those on the port engine bore both Junkers and Schwarz emblems! Note that the upper and lower surfaces demarcation line followed the centre line of the wing leading edge.

Right and lower. The Ju 88 V6 D-ASCY photographed in mid-1938 prior to the application of its paintwork. Usually major components were pre-sprayed prior to final assembly but in this case only the propeller spinners appear to have been painted.

In a display of potency, *Général* Vuillemain inspects the Bf 109C-1s of JG 132. The aircraft were camouflaged in 70/71/65 finish.

flage color for upper surfaces. (10R 2.5/6.)

62 — *grün* — (green) — used as a camouflage color for upper surfaces. (5GY 5.5/4.)

63 — *hellgrau* — (light grey) — used as an external finish for landplanes. Incorporated into later camouflage schemes as a color for upper surfaces. (10BG 7.2.)

65 — *hellblau* — (light blue) — used as a camouflage color for lower surfaces. (10BG 6/2.)

66 — *schwarzgrau* — (black-grey) — cockpit interiors. (N3 6.6%.)

70 — *schwarzgrün* — (black-green) — used as a camouflage color for upper surfaces on land-based aircraft. (10GY 2.5/1.)

71 — *dunkelgrün* — (dark green) — used as a camouflage color for upper surfaces on land-based aircraft. (5GY 3/1.)

72 — *grün* — (green) — used as a camouflage color for upper surfaces on aircraft operating over the sea. (5BG 3/1.)

73 — *grün* — (green) — used as a camouflage color for upper surfaces on aircraft operating over the sea. (10G 3/1.)

It will be seen from this list that there are many numbers missing from the sequence, these referring to colors used for internal applications. For example 7170.42 was stipulated for accumulators while instrument panels were to be finished with 7107.41. In both cases 41 and 42 were greys but no listing has been found which gives any precise indication of the exact shade of each. Many numbers in the sequence do not appear even in documents dating from early 1938 and it is assumed that they had possibly been already superseded or were never used.

Post-war analysis of these vital years has shown that both sides were still far from ready for a major conflict and many obsolescent types were readied for action by the simple expedient of painting them in camouflage. Thus the few ageing Ar 68Es remaining on front-line strength were soon seen bearing the new scheme. Military exercises were also a universal feature of this period and like the RAF, the *Luftwaffe* adopted a system of temporary markings to identify the 'enemy' force.

These temporary markings took the form of a black or red disc which obscured the wing and fuselage *Balkenkreuz* in all six positions. The paint used, 7120, was a glyptal resin-based temporary distemper and it was removed by using rubber scrapers. The surface was then cleaned and washed with petrol. This temporary finish had to be removed within eight days or it deteriorated and became difficult to erase.

Following the final annexation of the remainder of Czechoslovakia, the outbreak of hostilities became simply a matter of time. The *RLM* put this delay to very good use and an entirely new set of code markings was introduced for all front-line aircraft. Before this is detailed however, special mention must be made of the distinctive style of markings that had been carried by the Hs 123s. The normal five-character code was

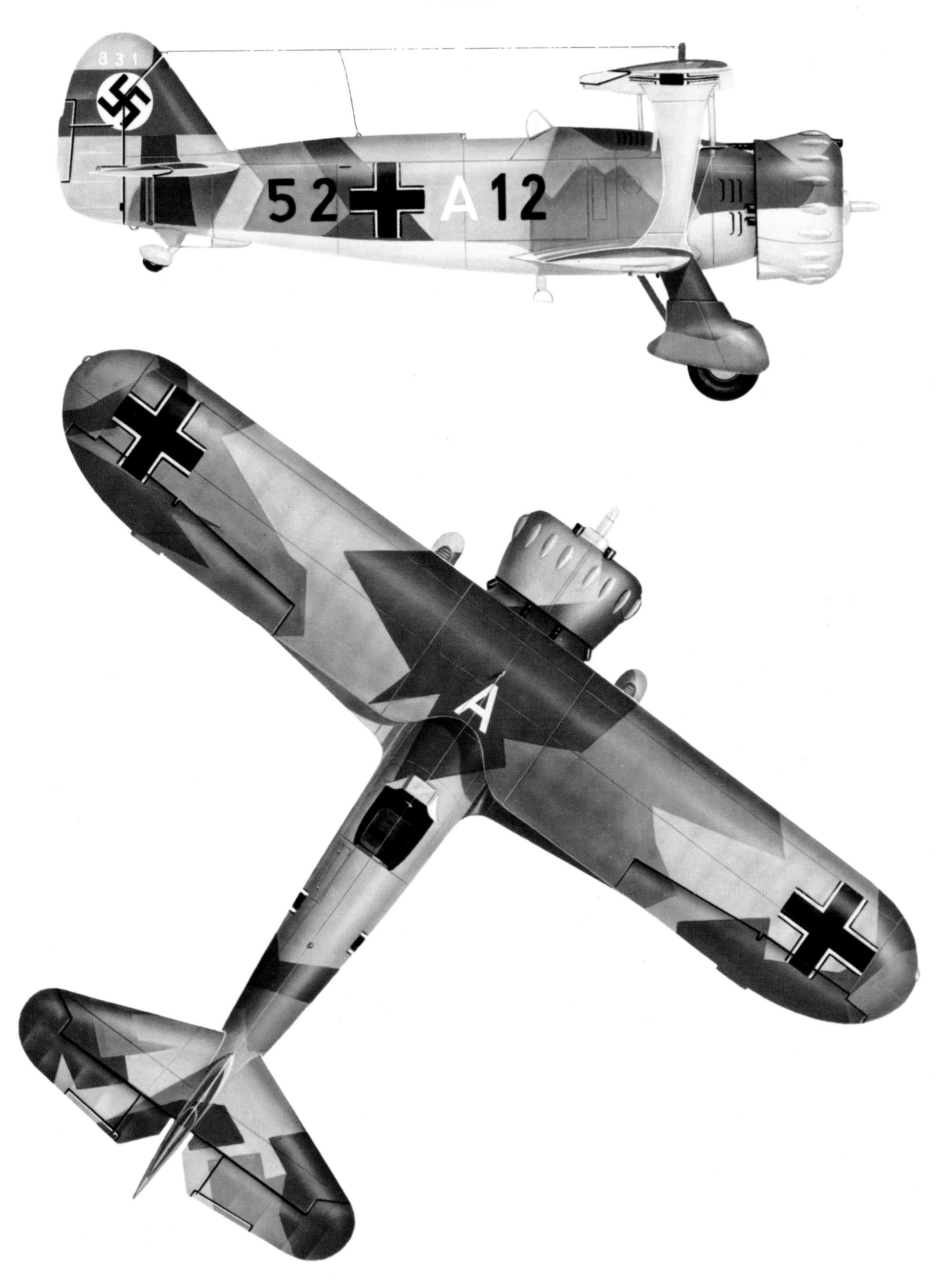

The Hs 123 was individualistic in that its wing upper surfaces carried only the aircraft's letter and not the entire code group as seen on other aircraft types. This particular aircraft, from 2 *Staffel* of I/St G 165, carried a 61/62/63/65 camouflage scheme but with pattern variations as explained in the main text. The narrow chord of the bottom wings and the restricted space made it necessary for the lower wing surface *Balkenkreuze* to be marked beneath the main wings.

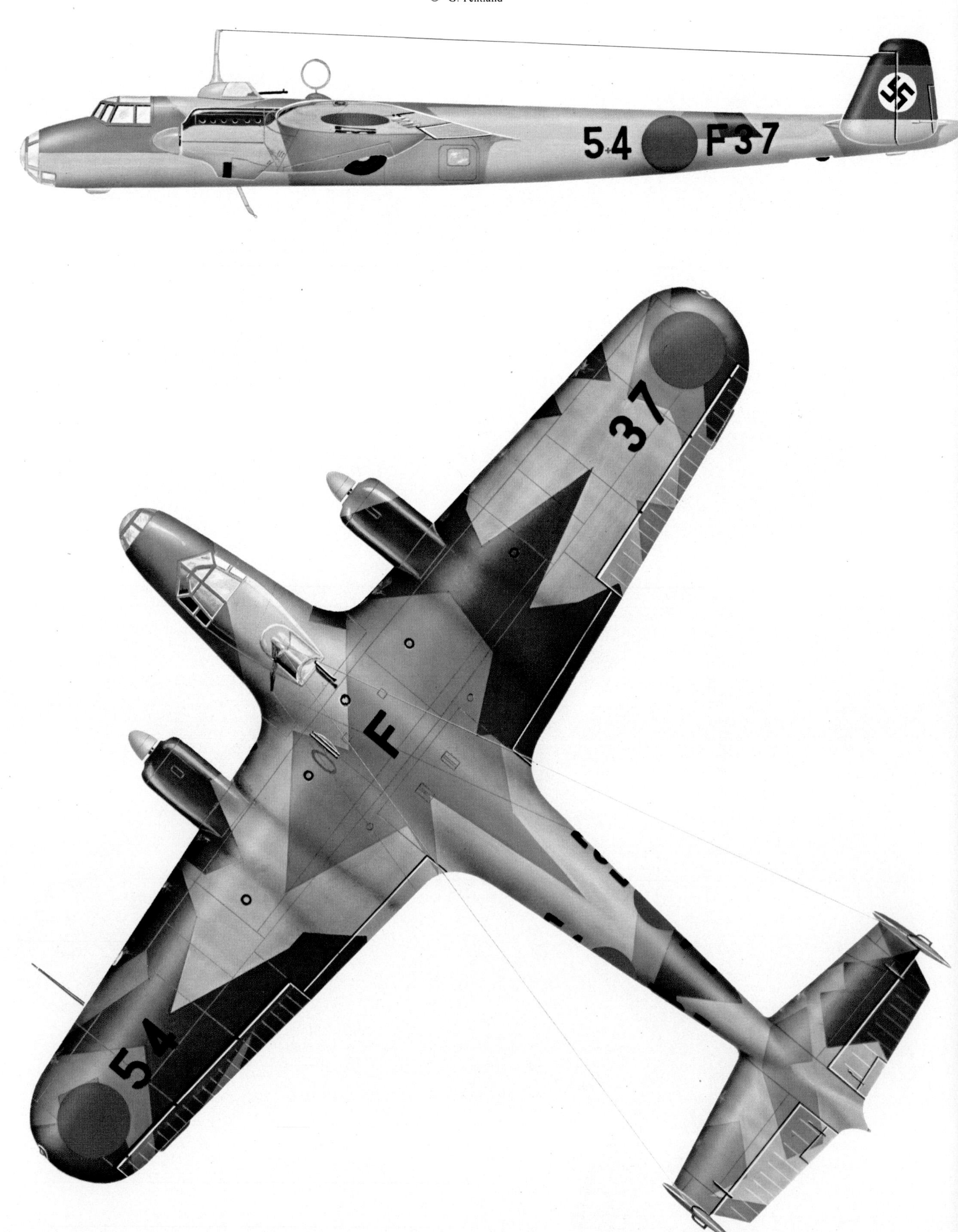

A Do 17E-1 of 7 *Staffel* from III/KG 255 displays the red disc markings used during the air exercises of 1938. The color scheme was standard 61/62/63/65, the shoulder-mounted wing of the Do 17 producing the characteristic high division line along the fuselage between the upper and lower surface colors. The letter F was repeated in black on the centre line of the ventral fuselage.

Part of the training school equipment at Schleissheim, this Fw 56 was finished in overall 63 and bore a single distinguishing band of white around its fuselage. Aircraft from other flights within this school carried different combinations of bands and colors as explained in the text. The civil registration was repeated beneath the wings.

A He 59 of the *DVS* at List in mid-1935. The aircraft carried the standard markings of this early period; black civil registration letters, the National Socialist emblem on the port side of the fin and rudder, with the national colors of black, white and red on the starboard side. The number on the fuselage side identified the aircraft within the school. With the exception of the floats—which were natural metal coated with a protective clear varnish—the entire aircraft was finished in the overall dark greenish-grey that was to become standard for seaplanes.

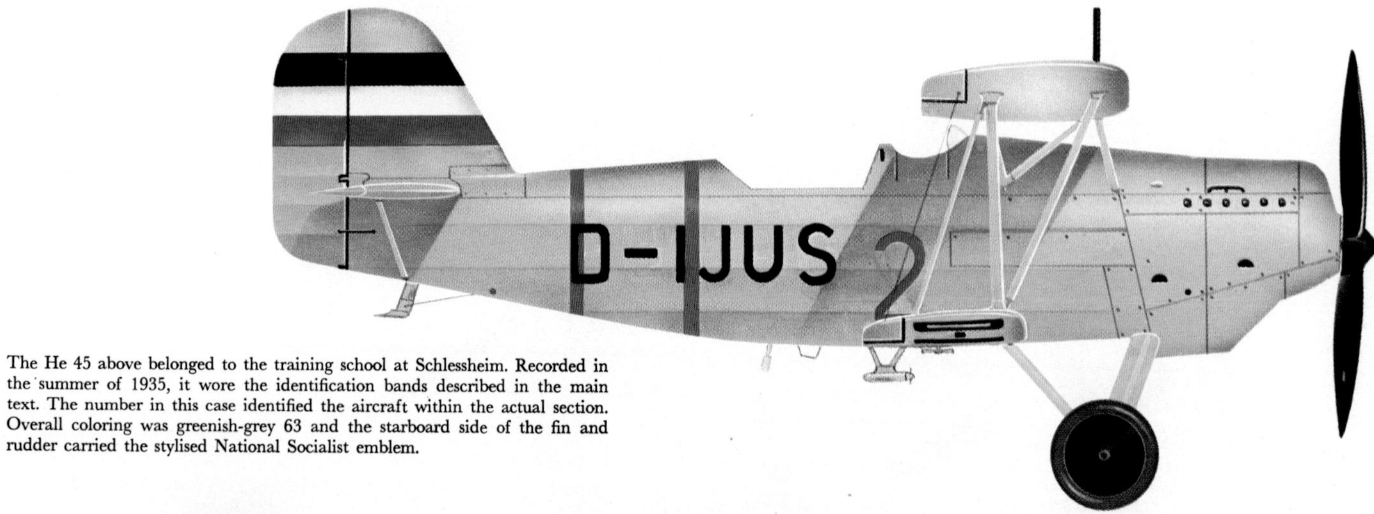

The He 45 above belonged to the training school at Schlessheim. Recorded in the summer of 1935, it wore the identification bands described in the main text. The number in this case identified the aircraft within the actual section. Overall coloring was greenish-grey 63 and the starboard side of the fin and rudder carried the stylised National Socialist emblem.

A Bf 109B of a *Jagdfliegerschule* of *Luftkreiskommando* II bearing the controversial markings as explained in the main text. Color scheme was standard for the period; 70/71/65. The *Hakenkreuz* remained overlapping both fin and rudder following the deletion of the red background and white disc.

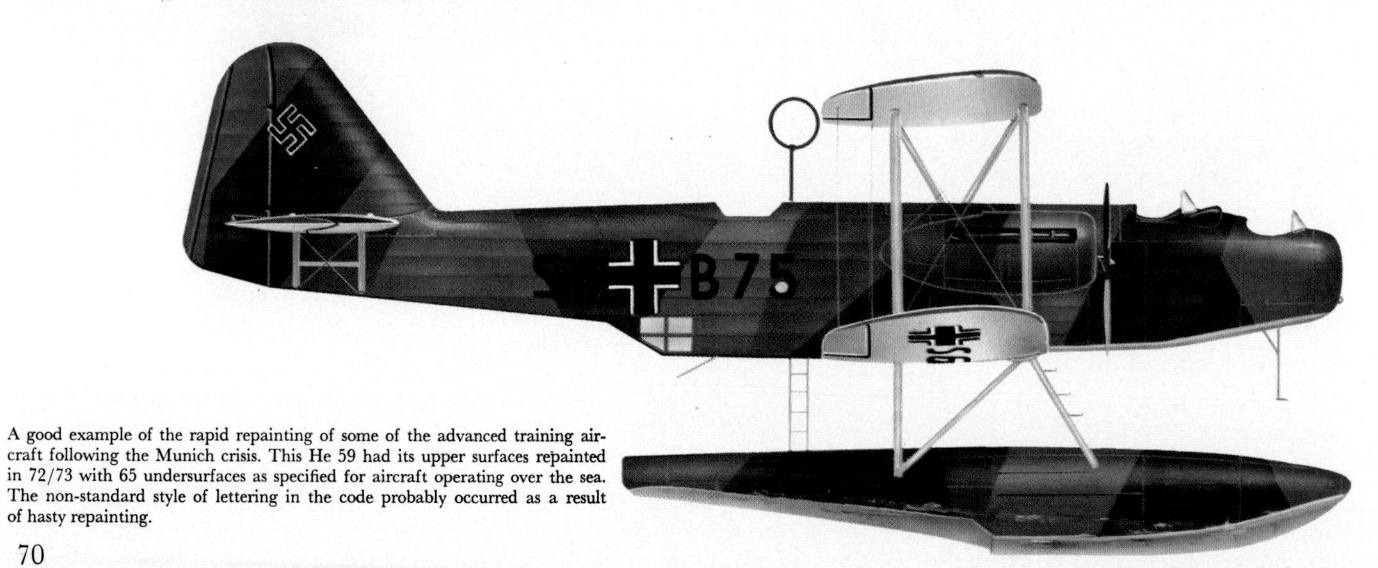

A good example of the rapid repainting of some of the advanced training aircraft following the Munich crisis. This He 59 had its upper surfaces repainted in 72/73 with 65 undersurfaces as specified for aircraft operating over the sea. The non-standard style of lettering in the code probably occurred as a result of hasty repainting.

retained on the fuselage sides but was deleted from beneath the wings and the upper surface of the top wings. Only the aircraft's own code letter was displayed on the upper surface of the wing centre section. In some instances this letter was applied in white, in which case the one each side of the fuselage was also painted white.

To return to the spring of 1939, both bomber and fighter code styles were replaced by a new, more colorful, more positive, and simplified system. This proved to be the last major revision implemented and the *Luftwaffe* used this method of identification right through until the close of hostilities.

Fighter codes remained a simple numerical sequence carried on the fuselage sides but the colors white, red, and yellow were introduced to distinguish individual *Staffeln* within a particular *Gruppe*. These were combined with the identification symbols as shown in the diagram on Page 77. As always there were exceptions to the rule and while the official directives dictated the placing of the aircraft's code number forward of fuselage *Balkenkreuze,* there were examples of it applied aft of the *Balkenkreuz,* coupled with the *Gruppe* identification symbol. The *Gruppe* symbol was generally applied in the *Staffel* color. Alternatively there were those who interpreted "forward of the fuselage *Balkenkreuz*" as meaning on the side of the engine cowling. This practice was to linger on for some time.

Bomber codes were more extensively modified, being reduced to a four-figure/letter combination divided equally either side of the fuselage *Balkenkreuz.* The first two, a figure and a letter, was the *Geschwader* code; the first letter after the *Balkenkreuz* was the aircraft's code letter within the *Staffel;* and the last letter identified the *Staffel* within the *Geschwader.* The aircraft's code letter was supposed to be painted in the *Staffel* color of either white, red, or yellow. Alternatively it could be painted in black and thinly edged in the appropriate color. In practice however, the early months after the new scheme was introduced saw many applications of the full code in black, or else the aircraft's code letter was simply edged thinly in white or painted completely in white regardless of the *Staffel* color requirement. The aircraft's code letter was intended to be applied beneath each wing and above each wing in

A group of Hs 126A-1s straight from the production line and camouflaged in 70/71/65. This color scheme was gradually being extended to other classes of aircraft and its use on new machines was accelerated by the Munich crisis. The near aircraft carried its *Werke Nummer,* 3058, in small white numerals, without the usual W Nr prefix, high up on the rear fuselage just forward of the fin. The use of civil registrations on factory-fresh aircraft had been superseded by 1938. Instead, a four-letter combination was employed utilising the company two-letter-type abbreviation. In this case the aircraft all carried four-letter codes commencing with the letters HS for Henschel. The nearer aircraft was thus coded HS under the starboard wing and IS, its production sequence code allocation, beneath the port wing. The white lines on the side of the fuselage were sighting lines for use with a hand-held aerial camera.

Upper. An interesting and somewhat confusing example of the art of subterfuge. German propaganda was as active as that of any other nation during the tense period of the late 1930s. The graceful He 100 V8, D-IDGH, captured the world air speed record with a speed of 463.92 mph, an achievement which was soon bettered by the Me 209 V1. Of greater importance to the propaganda machine however was the need to convince Allied Intelligence that the Heinkel product was actually in front-line service with the *Luftwaffe,* but the methods adopted were confusing and somewhat clumsy. The aircraft was painted in spurious and erroneous markings which should have provided some obvious clues to an astute observer of *Luftwaffe* camouflage and markings! The use of an obsolete fighter-style code that had not been used for more than two years was an error even further compounded by the incorrect grouping of the actual elements of the code. The color scheme of what appears to be overall 62 was also unusual.

Opposite, upper. A factory-fresh Ju 87B finished in 70/71/65. Like the previously illustrated Hs 126s, the aircraft carried a four-letter registration, the letters JU being visible beneath the starboard wing. As on the previously illustrated Ju 87, D-ILEX, the red band did not go right around the leading edge of the fin. Opposite, lower. An excellent example of the new 70/71/65 bomber camouflage scheme introduced in late-1938. This Do 17Z-1 shows the extreme outboard position of the wing *Balkenkreuze*. The chordwise yellow lines on the wings defined the areas beyond which personnel were not permitted to walk. The small legends outboard of these 40-mm wide lines read *Nicht Betreten* in 50-mm high letters.

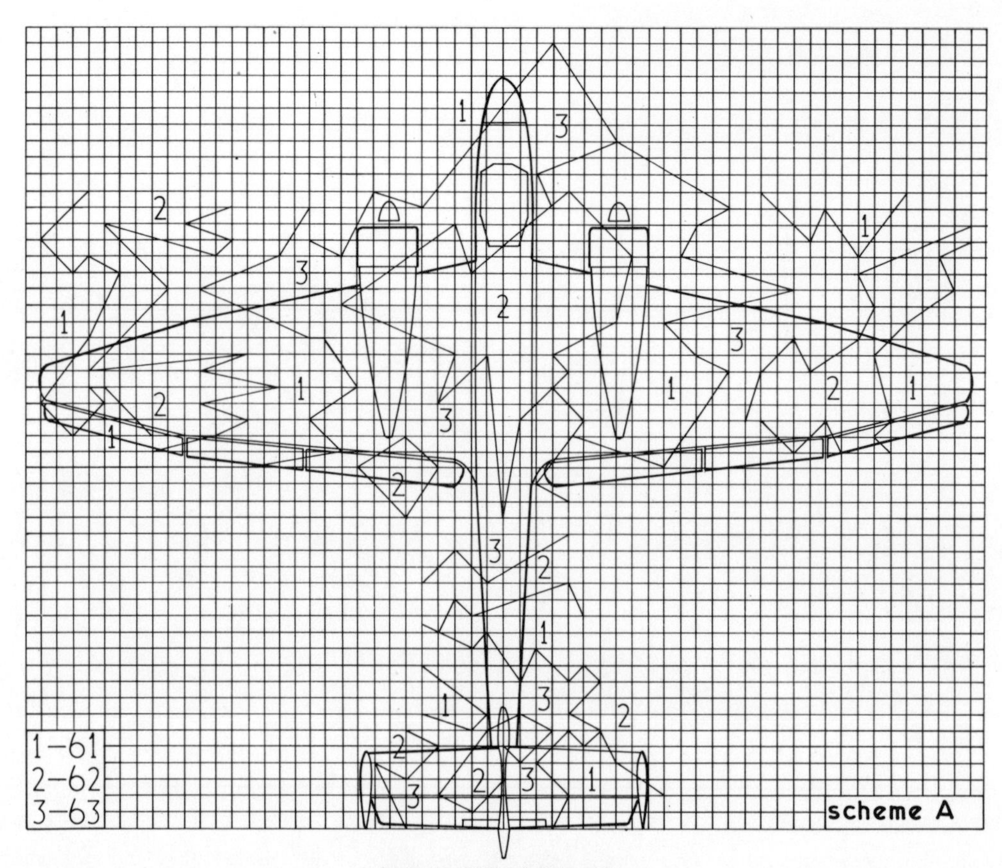

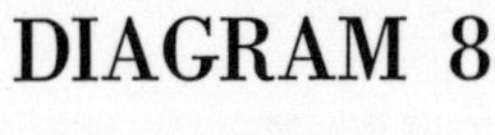

DIAGRAM 8

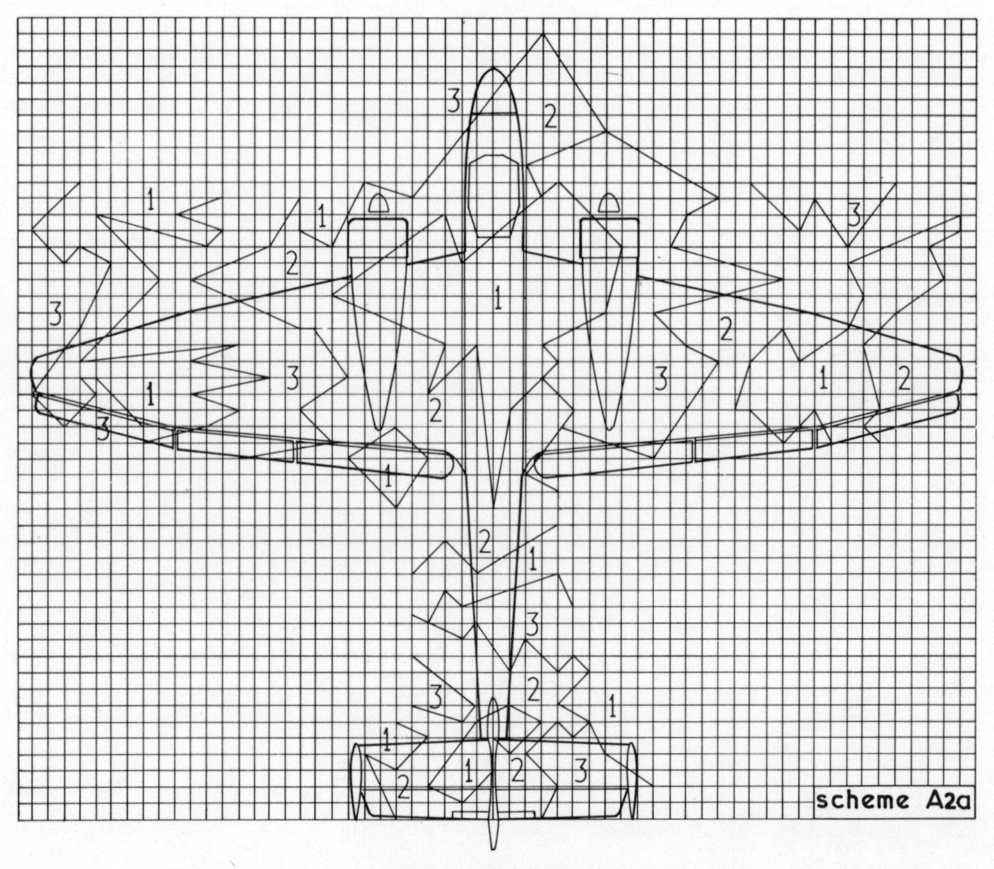

DIAGRAM 9

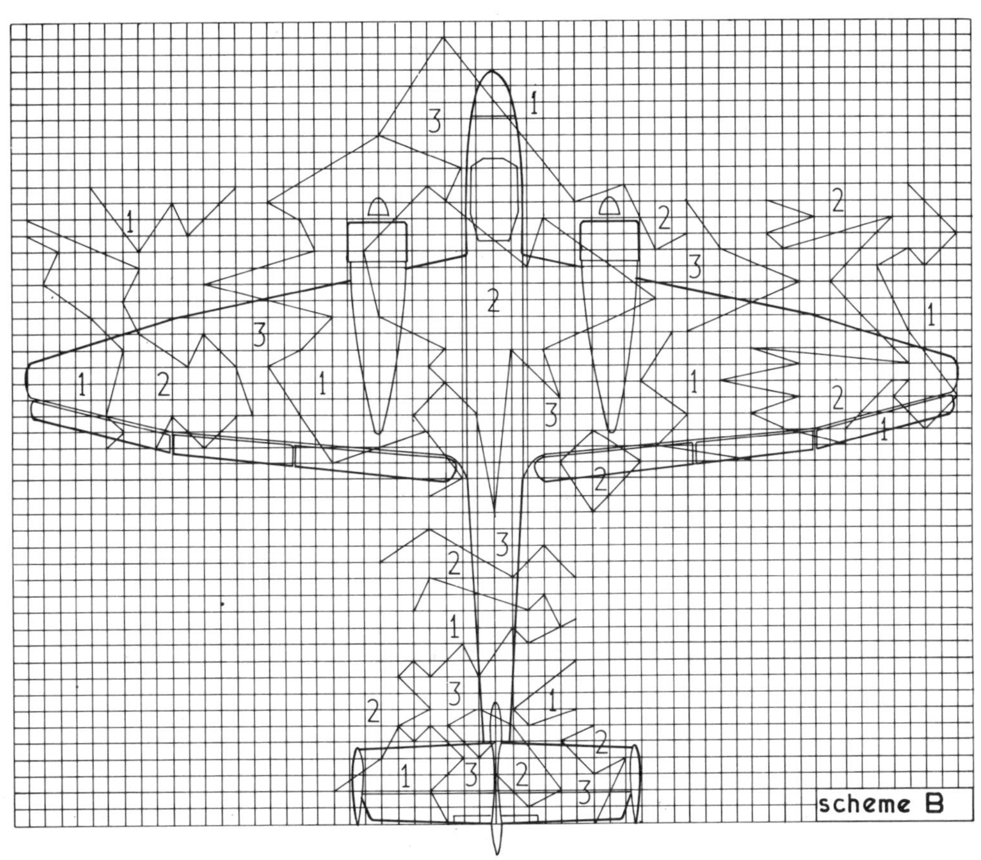

DIAGRAM 10

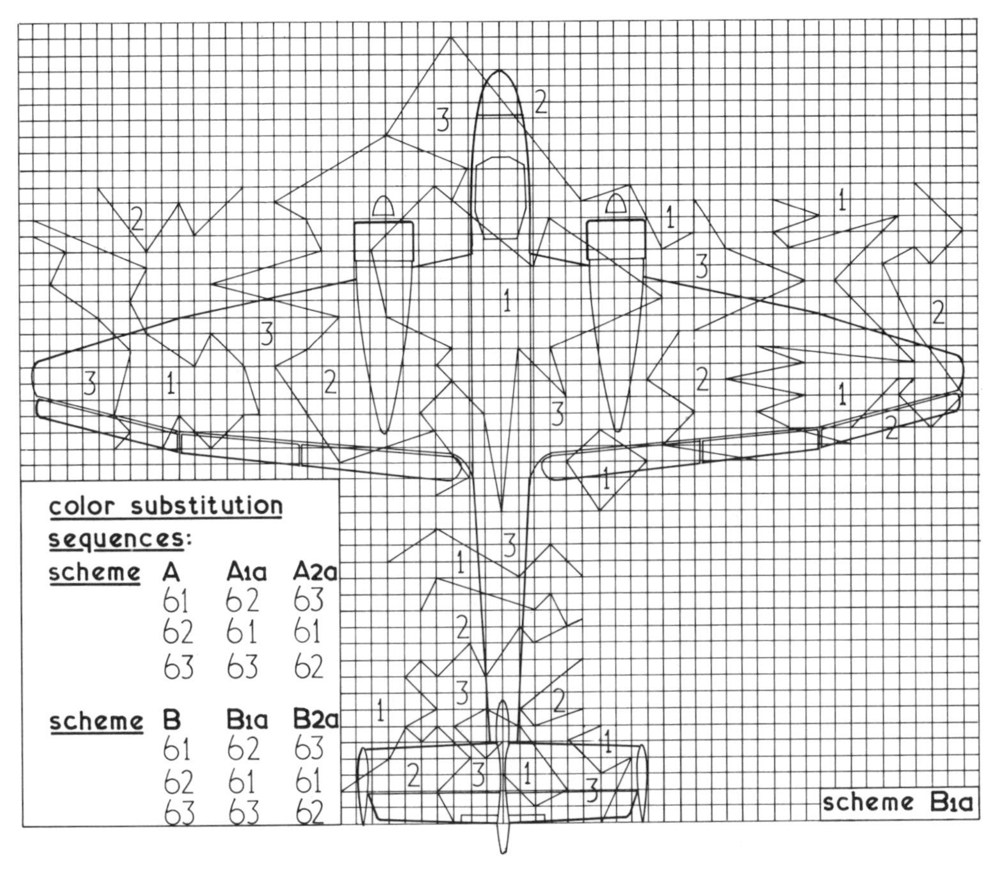

DIAGRAM 11

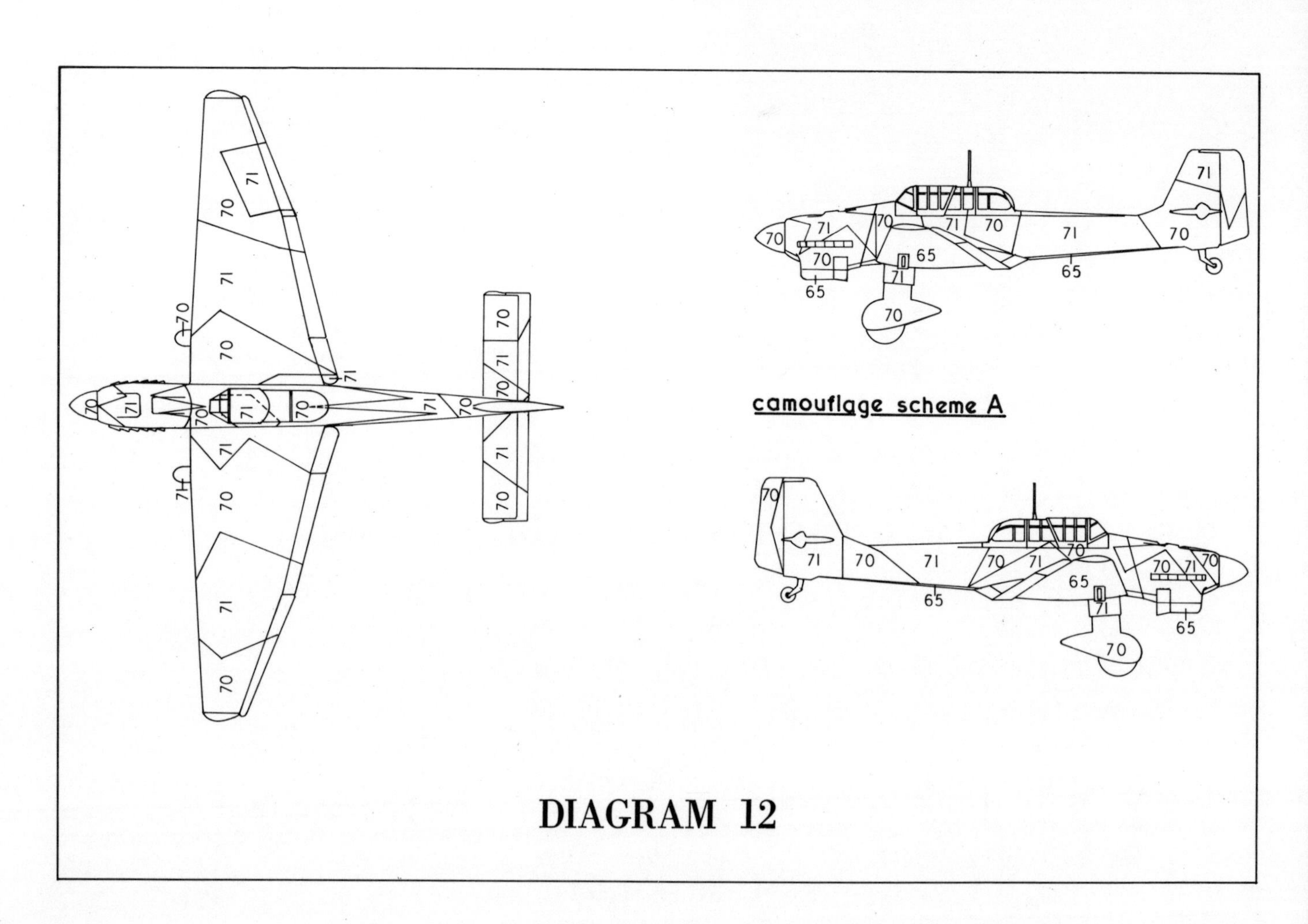

DIAGRAM 12

DIAGRAM 13

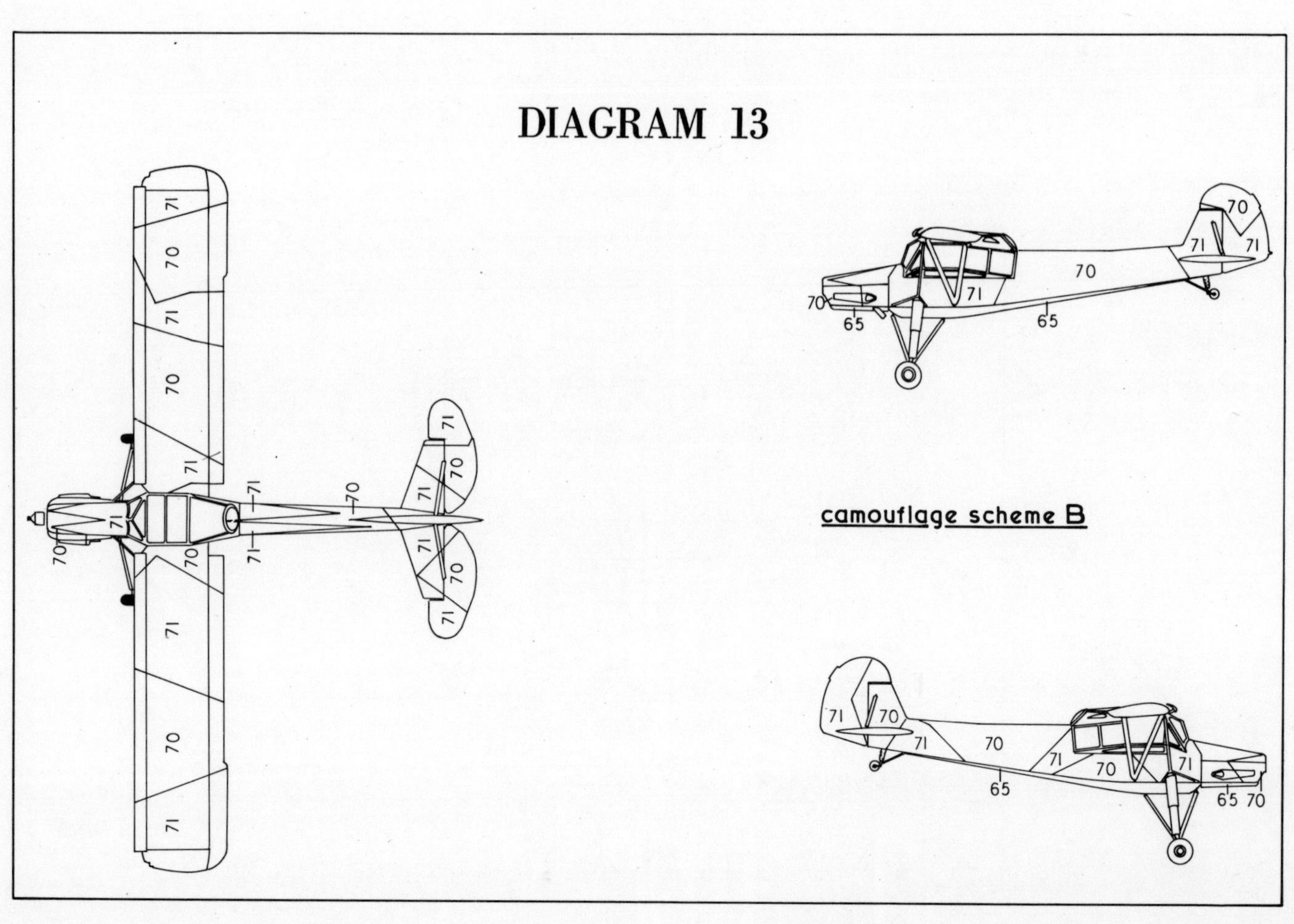

JAGDVERBÄNDE					
STAB	GESCHWADER SYMBOL				GRUPPE SYMBOL
KOMMODORE	≪| OR ⋘|				≪ OR ⋘
ADJUTANT	<|				<
IA	<– OR ←–				NONE
TO	<|O				<O
MAJOR	<||				NONE
GRUPPE	I	II	III	IV	STAFFEL COLOR
STAFFEL	1	4	7	10	WHITE
STAFFEL	2	5	8	11	RED
STAFFEL	3	6	9	12	YELLOW
GRUPPE SYMBOL	NONE	—	~	+ OR ●	

DIAGRAM 14

KAMPFVERBÄNDE						
STAB	LAST LETTER OF CODE		AIRCRAFT LETTER COLOR			
GESCHWADER	A		BLUE			
GRUPPE I	B		GREEN			
GRUPPE II	C		GREEN			
GRUPPE III	D		GREEN			
GRUPPE IV	E		GREEN			
GRUPPE V	F		GREEN			
GRUPPE COLOR	I WHITE	II RED	III YELLOW	IV BLUE	V	STAFFEL COLOR
STAFFEL	1	4	7	10	13	WHITE
LAST LETTER	H	M	R	U	X	
STAFFEL	2	5	8	11	14	RED
LAST LETTER	K	N	S	V	Y	
STAFFEL	3	6	9	12	15	YELLOW
LAST LETTER	L	P	T	W	Z	

black, inboard of the respective *Balkenkreuz*. Old habits die hard however, and many units applied the entire four-figure/letter code to both upper and lower wing surfaces; ie a figure and a letter inboard of one *Balkenkreuz,* with the other two letters inboard of the other *Balkenkreuz*. Time and patience would eventually remedy many of these anomolies but those first months provided an interesting multitude of variations.

With the exception of the letters G, I, J, O and Q the remainder of the alphabet was utilised for *Staffel* identification codes. Reference to the diagram on Page 77 will clarify this system. Although this produced sufficient code allocations for five *Gruppen,* the full range was rarely, if ever, employed, due to the vast number of aircraft involved. Each *Gruppe* was allocated a color; white, red, yellow, and blue respectively, which was also applied to the propeller spinner and often in combination with the *Staffel* color, the latter being applied to the tip area. A few bomber units adopted a colored nose band to identify the *Gruppe*, but the practice was not widespread.

A new class of aircraft, the long-range heavy fighter, entered service in 1939 in the form of the Bf 110. It conformed to the general standard as far as camouflage and national markings were concerned but was unique regarding codes and special markings. It utilised bomber-style identity codes but combined with them the use of *Geschwader* and *Gruppe Stab* symbols. The practice of painting the propeller spinners in either the *Staffel* color or the combined *Staffel/Gruppe* color was not extensive at this early stage. Like the single-engined fighters, spinner colors were not common as yet, but the first year of war was to see a distinct change.

The art of the 1914-18 heraldry had gradually reappeared and was in almost universal use in the *Luftwaffe* by the outbreak of war. While not officially encouraged, the practice was certainly not discouraged either. Unlike the old brigade however, the new generation's heraldry was confined to emblems of reasonable proportions and it certainly never became general practice to paint entire aircraft in gaudy stripes or monochromes of violent hue.

Despite this, color and imagination was not lacking. The *RLM*'s policy of rotating crews through Spain for experience, then appointing the more successful officers to senior positions in home-based units, added to the scope of the emblems and devices.

It should be noted that the system of allocation of emblems was quite a complex affair and certainly with the larger size of the bomber units, it was to increase in complexity. Theoretically, there could be as many as thirteen distinct emblems in use in a basic *Geschwader* of three *Gruppen;* the *Geschwader* emblem, the three *Gruppen* emblems and nine *Staffel* emblems. In practice some of the units utilised a system whereby the actual emblem remained constant but the background varied in color to match the *Gruppe* color. This practice was further enhanced by the addition of personal emblems.

And so the *Luftwaffe,* proudly displaying its new heraldry, stood poised for a war that was now plainly inevitable.

He 111Ps displaying their white spinner markings. The nearest aircraft was coded 25 + B33, a style of coding for heavy aircraft that was soon to be replaced by a simpler and more positive system.

Prague, March 1939. *Generaloberst* Ernst Udet, accompanied by German and Czechoslovakian officers, examines the *Luftwaffe's* acquisitions following the occupation of the remainder of Czechoslovakia. The aircraft had been incorrectly and no doubt hastily marked with a *Hakenkreuz* in all six positions.

A Ju 87B awaiting delivery from the manufacturers. The color scheme was 70/71/65 and the undersurfaces of the wings bore the letters JU-EK. This aircraft was the 29th in this production sub-type series as indicated by the large white 29 painted on the rear fuselage. The red background and white disc of the old *Hakenkreuz* emblem had not been applied in accordance with the regulations introduced after the Munich crisis. The propeller blades were also painted to derive maximum benefit from the camouflage.

The 25th production Bf 110C-1 displaying markings similar to those on the Ju 87B previously illustrated. The camouflage colors were identical but its production number had been applied in a different location but still in white. Note also the lack of any form of four-letter code starting with the company-type initials. The practice was phased out fairly rapidly during this period.

Bf 109E W Nr 878 shows its very dark 70/71 upper surface camouflage. The *Hakenkreuz* remained, spanning both fin and rudder with only the thin regulation white surround. The chevron marking denoted that it was the mount of the *Gruppe* adjutant, while the horizontal bar signified II *Gruppe*. The small figure 2 marked the actual physical status of the pilot and aircraft within the *Gruppenstab* structure. The canvas cockpit cover bore the aircraft's type identification and *Werke Nummer* in white, a peacetime luxury that was to become gradually rarer as the war progressed.

Other aircraft also wore quasi-civilian markings in conjunction with military ones. This Ju 52, respendent in overall 63 with black trim, bore full civil registration with a *Balkenkreuz* separating the D-prefix from the remainder. The lack of any red background to the *Hakenkreuz* was a further indication of its military ownership. It is believed to have been operated by the *Luftfeldpost Staffel* responsible for the delivery of military mail.

A Do 17E of the *Stab* flight of KG 77, 3Z + DA, on the eve of war. The aircraft retained the old 61/62/63/65 camouflage scheme but the red band and white disc background to the *Hakenkreuz* had been overpainted in 61. This color was generally used for this purpose as its dark pigmentation provided the best opaque effect over the red underneath. Like many bomber aircraft of this period, the new unit codes had been incorrectly applied. The letter D should have been blue and not black to indicate that the aircraft belonged to the *Geschwaderstab*. In common with other aircraft of this particular *Kampfgeschwader* however, a band had been painted around the nose of the aircraft in the appropriate blue color. The propeller spinners were also blue. For aircraft of other than the *Stab* flight, the spinners were painted in the *Staffel* color and the nose band in the *Gruppe* color. The small wooden sign leaning against the port wheel carried the last two letters of the code, DA.

CHAPTER 4

TRAINING FOR PERFECTION

The origins of the *Luftwaffe's* training scheme were of necessity as clandestine as were its own beginnings. Instilled with a firm resolve to restore and expand a military air arm, the architects of the new *Luftwaffe* found it necessary to provide an organization capable of supporting such a force. By devious methods a training organization was established under civilian guise and thus the *DVS (Deutsche Verkehrsfliegerschulen)*, the German Commercial Pilots' School, provided the means to train pilots while the German aircraft industry was given the incentive to design, develop, and build the aircraft.

By 1932, the Heinkel company had developed a very useful two-seat design, the He 45, which was to fulfil a number of important training rôles. Initial deliveries to the *DVS* began in the summer of 1933, the majority going to the school at Schliessheim. These machines bore standard civil registration markings, eg D-2245, in black on the wings and fuselage. The overall finish of the aircraft consisted of either clear lacquer or aluminium dope over natural linen fabric. The He 45s were also supplemented by small numbers of another of the firm's products, the He 50, the first of these arriving late the same year.

Advanced training techniques in navigation and night flying were undertaken in Dornier Do 17Gs of the *Deutsche Reichsbahn* (German State Railway) which conveniently began operating an air freight service in November 1933. For this training, pupils were seconded from the *Verkehrsinspektion der DLH* (Traffic Inspectorate of *Deutsche Luft Hansa)* which was, in reality, the newly-established *Behelfsbombergeschwader* (Auxiliary Bomb Group).

Training equipment was almost exclusively composed of Heinkel products and the *DVS* seaplane training school at List received the first supplies of the He 59 twin-engined floatplane in 1934. These were supplemented by single-engined He 60s.

It was obvious that while the nucleus of World War I pilots lasted, the needs for *ab initio* training would remain small, but eventually this resource would inevitably exhaust itself. Thus arose the need that was fulfilled by the Arado Ar 66 and the Bücker Bü 131 in 1934. The *DLV (Deutscherluftsportverband)*, was the main recipient of these and thus a constant stream of new pilot material was assured.

With the public declaration of the *Luftwaffe* in March 1935, the various flying training schools were able to reveal the military nature of their task. Apart from their civil registration markings however, and the National Socialist fin and rudder markings, they carried no other marking by which the individual school's identity could be established. Division of the aircraft into sections or flights within a training school was done visually. The school at Schleissheim, near Munich, used a system of colored bands applied around the full girth of the fuselage aft of the wings. Single bands of red, white and yellow, and also double bands of each color are known to have been used to identify the various sections.

The introduction of the new military codes in 1936 was naturally extended to include the training establishments. Although superficially similar in appearance to the codes used by front-line units, there were certain built-in and quite subtle distinctions. The most prominent of these was the substitution of the letter S for the first numeral of the sequence. The other variations were not quite so apparent and are best explained by the use of an example; eg Arado Ar 66 coded S2 + V12.

S — *Schule* (School) aircraft.
2 — *Luftkreiskommando* II, Berlin.
V — Flight identification letter within the school.
12 — Numerical identity of the aircraft within the particular school.

It is to be noted that the third symbol no longer identified the individual aircraft, but identified the flight to which the aircraft belonged within the school. The last two digits were grouped together to provide the aircraft's personal identification. This accounts for the relatively high numerical sequences sometimes seen, eg two Bf 109Bs coded S2 + M57 and S2 + M58. Thus they were the 57th and 58th aircraft within the school and both belonged to the same flight. Such a system was both flexible and adequate without compromising the actual identity of the school. Little information has survived concerning the administration, size, and composition of the early schools of this period and thus it is not possible to make a general identification of which flight identification letters were allocated to the various schools. From the evidence that has been assembled however, it would appear that flight identification letters were issued alphabetically to the various schools within a particular *Luftkreis*. For example if there were five schools within a particular *Luftkreis* then, say, letters A to D were allocated to the first one, E to G to the next one, and so on.

Quite obviously the number of flights varied from school to school and possibly for security reasons certain blocks of letters may not have been allocated in one *Luftkreis* or another. The seeming possibility of confusion between the aircraft of one *Luftkreis* and another would not actually occur due to the second character of the code sequence; eg S2 + B15 and S6 + B09. Both aircraft belonged to B Flight but the first was

A brand new Ar 66 *ab initio* training aircraft prior to the application of its civil registration. The scheme was silver colored aluminium dope with simple black trim around the exhaust ports.

A silver colored Fw 44E *Stieglitz*. The fuselage upper decking from the cockpit aft was painted black as were the wheel hubs. The civil registration, D-2621, was the style used between mid-1933 and Spring 1934. The starboard side of the fin and rudder carried the three horizontal stripes of black, white and red.

Another early model Fw 44, a type which was to bear the brunt of primary training for many years. Finished in a similar scheme to D-2621, the style of its registration dates the period as pre-Spring 1934.

operated by a school in *Luftkreis* II (Berlin) and the second by a school in *Luftkreis* VI (Kiel).

In other publications dealing with this aspect of *Luftwaffe* markings, it has been stated that this type of coding was applied only to aircraft of the *C-Schulen* (advanced training schools) in general and bomber-reconnaissance training aircraft in particular. Photographic evidence would appear to refute this conclusion since, for instance, Arado Ar 66, coded S2 + V12, was certainly not the type of machine used at such advanced schools. The Bf 109B aircraft of the *Jagdfliegerschule* mentioned above similarly do not fall within the prescribed category yet they and others like them also bore the full S-type codes. Certainly from 1936 onward the newer aircraft of the *A/B-Schulen* (elementary training schools) appear to have utilised only the standard civil registration until January 1939, when the coding system for all second-line aircraft was revised. From that date onward, all training aircraft bearing civil registration markings had the prefix letter D deleted and the letters WL substituted. The latter stood for *Wehrmacht Luft* and identified the aircraft belonging to the armed forces as opposed to civil aircraft which still retained the prefix letter D.

The institution of this order produced a variety of code applications and modifications. New aircraft leaving the various factories were marked in accordance with the requirements of the new order; ie the prefix letters WL separated from the remainder of the code by a standard-proportioned *Balkenkreuz*. Retrospective modifications to existing machines however, produced the spate of variations mentioned. In some instances the civil registration letters had been applied as large as possible on the fuselage sides. Thus when the letter D was deleted, it left relatively little room for the application of the letters WL. The shape of the L allowed some degree of saving grace as it was possible to interlock it with the *Balkenkreuz,* but the result was still somewhat crowded. It will be realised that as the *Balkenkreuz* was centred on, or rather, *over* the existing horizontal dividing bar of the registration, it was well forward on the port side and well aft on the starboard. In some instances the singular lack of room resulted in the horizontal black bar being retained in place of the specified *Balkenkreuz.*

Wing surface registrations were similarly modified by the substitution of the letters WL, but the dividing bar was retained and the *Balkenkreuze* were applied outboard near each wingtip.

The Munich crisis repercussions had not effected front-line aircraft markings alone. Equipment of the *Jagdfliegerschulen,* where potential fighter pilots received their advanced training, usually comprised obsolescent fighter types such as the He 51. These aircraft, like other obsolescent types still equipping some front-line units, were hastily repainted in the full 70/71/65 camouflage scheme. In some cases the segmented pattern of the upper surfaces was somewhat individualistic in its application. To distinguish their training rôle however, a broad white band was applied to the rear fuselage just forward of the tailplane. On this encircling band the aircraft's identity within the particular school was applied in black numerals. The top of the upper wing centre section was also painted white and the number was repeated, again in black, on this distinguishing panel. Standard *Balkenkreuze* appeared in all six positions and in common with other military aircraft of the period, the red background to the *Hakenkreuz* was deleted, along with the white disc, leaving only a thin white edging.

During the final annexation of Czechoslovakia, large quantities of indigenous Avia B-534 biplane fighters were seized. A few saw limited service with JG 71, but the majority were employed for advanced fighter train-

A great mainstay of the training organizations was the Bü 131 used by the *A/B Schulen.* D-EVEO provides an example of the revised civil registration style introduced in April 1934. The colors and markings were consistent with standards prevailing at that time. Notice the practice of painting the interplane struts black. As a general rule training aircraft allotted to these duties were usually painted in overall silver with a combination of any or all of the following; black wheel hubs, black interplane and tail struts, and black upper fuselage decking from the cockpit aft.

Upper. Wearing a class-A2 civil registration, this He 72 *Kadett* displays the standard silver colored aluminium doped finish with markings appropriate to the time. The photograph was taken on 3rd August 1935, six months after its last major inspection as recorded on the legend just forward of the tailplane. It gave the following information:

REKLAME-STAFFEL MITTELDEUTSCHLAND
DES D L V DOBERITZ

LEERGEWICHT	535 kg
ZULADUNG	325 kg
FLUGGEWICHT	660 kg
PERSONENZAHL	2
LETZTE NACHPRÜFUNG	20.2.35
NÄCHSTE NACHPRÜFUNG	20.2.36

Just above the red stripe across the fin appeared the legend W Nr 322 in small black letters and numerals.

The *Stösser* was originally designed to meet the requirements of a specification for a *Heimatschutzjäger* (home defence fighter). It was to serve a different but equally important rôle however, as an advanced fighter trainer, forming the mainstay of the *Jagdfliegerschulen* until 1945. Photographed not long after the He 72 D-ELIF, this lineup of five Fw 56s again shows the consistency of markings for training aircraft; overall silver finish with black wheel hubs and wing and tailplane struts. The propeller blades were also painted black and bore the manufacturer's name, HEINE, in what appears to be yellow block letters near each blade tip. The legend near the tailplane gave similar information to that listed for D-ELIF but with minor variations:

LEERGEWICHT: 670 kg
GESAMTLAST: 315 kg
FLUGGEWICHT: 985 kg
HÖCHSTZUL. PERS. ZAHL. 1
LETZE PRÜF. 20.8.35
NÄCHSTE PRÜF. 20.8.36
EIGENT. FLIEGERGRUPPE
HALTER: DAMM

The Fw 58B *Weihe* was used for a variety of training rôles; gunnery, navigation, and multi-engined pilot instruction at the *C-Schulen*. D-APAN carried a class C registration plus the usual fin and rudder markings. The overall color was 63, transition from the old 01 schemes taking place around mid-1935. Retrospective painting of older aircraft was very slow and in some cases was not carried out until after the outbreak of war when an official *RLM* order forbade the use of 01 as an external color.

A Do 23G bearing a seemingly unusual combination of markings. During the period between March 1935 and the introduction of military codes in 1936, aircraft of the *C-Schulen* utilised a combination of civil registrations and military markings. The positioning of the *Balkenkreuze,* which lacked any black edging, conformed rigidly to the current *RLM* instructions for military aircraft. Closer examination of this aircraft and similarly marked Do 23Gs reveals that the civil registration had also been applied in the proportions laid down for military codes, ie six tenths the height of the *Balkenkreuz.*

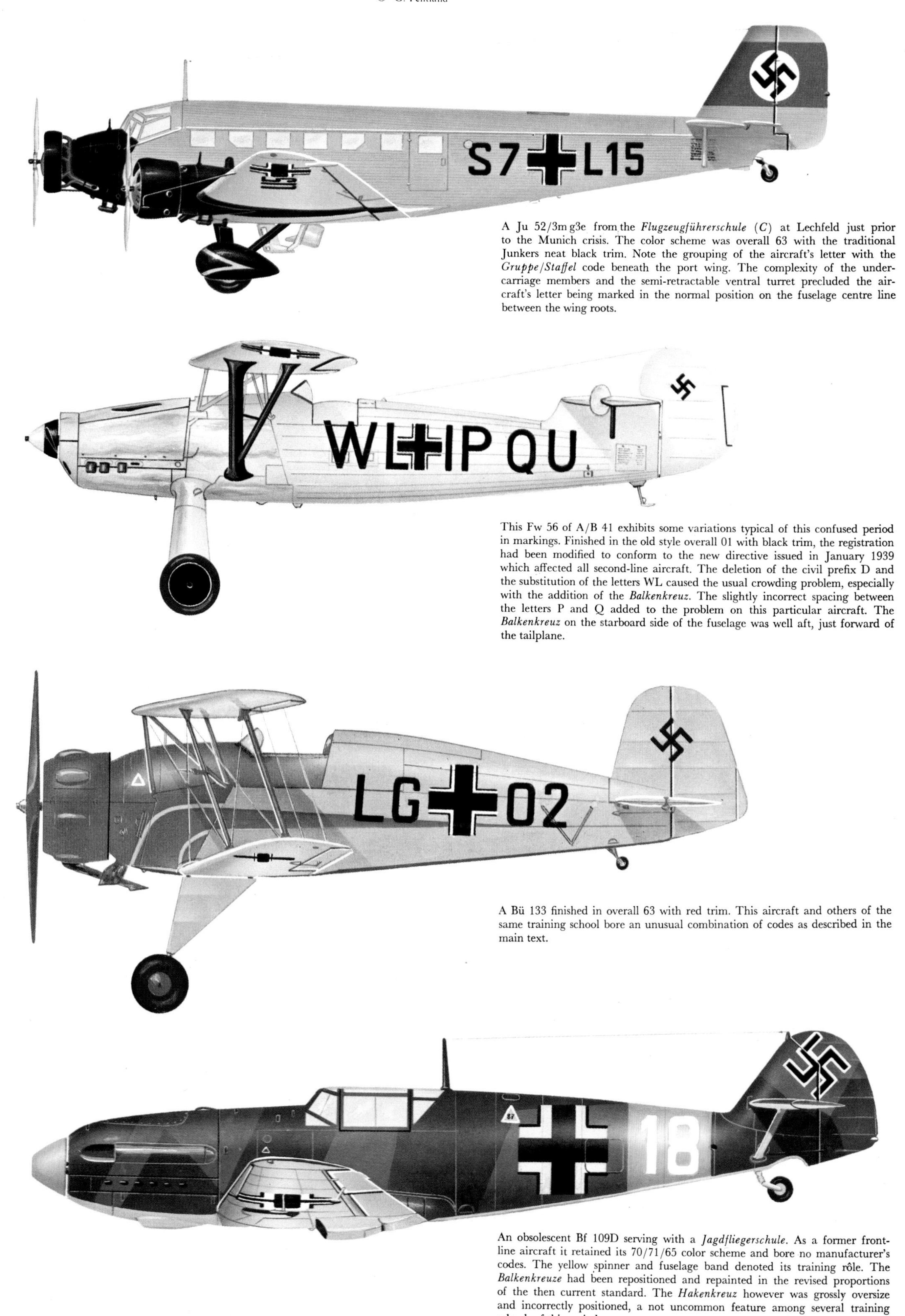

A Ju 52/3m g3e from the *Flugzeugführerschule* (*C*) at Lechfeld just prior to the Munich crisis. The color scheme was overall 63 with the traditional Junkers neat black trim. Note the grouping of the aircraft's letter with the *Gruppe/Staffel* code beneath the port wing. The complexity of the undercarriage members and the semi-retractable ventral turret precluded the aircraft's letter being marked in the normal position on the fuselage centre line between the wing roots.

This Fw 56 of A/B 41 exhibits some variations typical of this confused period in markings. Finished in the old style overall 01 with black trim, the registration had been modified to conform to the new directive issued in January 1939 which affected all second-line aircraft. The deletion of the civil prefix D and the substitution of the letters WL caused the usual crowding problem, especially with the addition of the *Balkenkreuz*. The slightly incorrect spacing between the letters P and Q added to the problem on this particular aircraft. The *Balkenkreuz* on the starboard side of the fuselage was well aft, just forward of the tailplane.

A Bü 133 finished in overall 63 with red trim. This aircraft and others of the same training school bore an unusual combination of codes as described in the main text.

An obsolescent Bf 109D serving with a *Jagdfliegerschule*. As a former front-line aircraft it retained its 70/71/65 color scheme and bore no manufacturer's codes. The yellow spinner and fuselage band denoted its training rôle. The *Balkenkreuze* had been repositioned and repainted in the revised proportions of the then current standard. The *Hakenkreuz* however was grossly oversize and incorrectly positioned, a not uncommon feature among several training schools of this period.

If practicable, captured operational aircraft were utilised by the *Luftwaffe* to supplement the equipment of the advanced training schools. This Avia B-534 retained its original color scheme of dark green upper surfaces and pale blue undersides but both the elevators and the rudder displayed a slightly lighter shade of green. Possibly these were replacement components that were painted with a rough color match. This seems to be borne out by the fact that the same color was used to paint out the original Czechoslovak markings on the upper surfaces of the top wings. The application of the *Hakenkreuze* to the rudder was unusual.

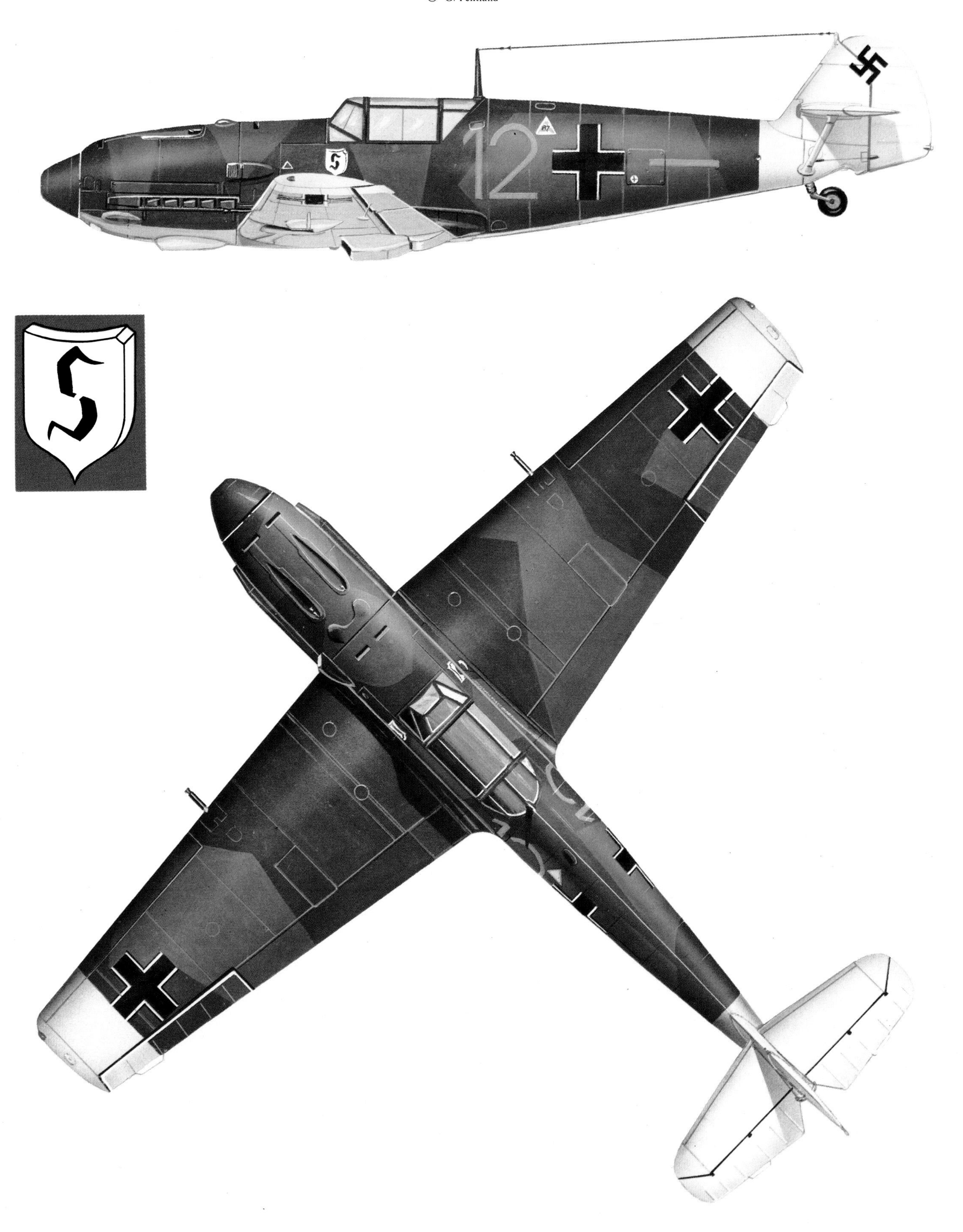

A Bf 109E of II/JG 26 as recorded in its condition during the military manoeuvres of August 1939. The bold white markings, a temporary feature only, were removed at the end of the exercises. The aircraft's number was correctly applied in the 6 *Staffel* color of yellow, but like many fighter aircraft of the 1938/39 period, the digits were too large, being the same height as the *Balkenkreuz*. Color scheme was standard 70/71/65, the *Geschwader* emblem of a stylized letter S appearing on both sides of the fuselage.

Camouflaged in 70/71/65, this Ju 87B of I/St G 1 shows the state of markings seen on many aircraft during the last days of the uneasy peace. The letter D on the fuselage had the appearance of being somewhat hastily repainted in the white *Staffel* color. It was the aircraft of this *Gruppe* which carried out the opening air attacks of the war. The diving raven emblem appeared on both sides of the fuselage.

He 60s of *Flugzeugführerschule* (*See*) Stettin, showing the style of codes introduced in 1936. Although of poor quality, this photograph helps to illustrate the point that the single letter to the right of the fuselage *Balkenkreuz* did have some significance, namely to identify the flight to which the aircraft belonged. On the left is S6 + D54 of D Flight, while the other two, S6 + B09 and S6 + B43, both belonged to B Flight. The overall finish of the aircraft was 02 with the floats painted in 01.

A closer look at another He 60, this time S6 + B05 of B Flight. On the original print it was possible to see the last three elements of the code, B05, beneath the bottom port wing. The necessity to group the flight letter with the aircraft number on machines with complicated undercarriage structures or floats is clearly demonstrated. If the letter B had been placed in the more normal position on the bottom of the fuselage between the wing roots, it would have obviously been difficult to see. The extremely dark, almost black, protective paint finish on the underside of the floats is clearly visible.

A Fw 58B bearing the WL code prefix in force between January 1939 and March 1940. The black code WL + OMGQ is almost lost against the dark 70/71 upper surfaces which show a considerable degree of highlight reflection. The *Balkenkreuze* again exhibit incorrect variations typical of this period. They had been modified by simply widening the white border to the new overall dimensions specified for fuselage and underwing *Balkenkreuze*. In doing this the thin black edging had been deleted. In addition, the *Balkenkreuze* on the wing upper surfaces had not only been left in their old extreme wingtip position but also, incorrectly, been given wide white borders. The *Hakenkreuz* was of the correct style, size and proportions but remained in the old central position overlapping fin and rudder. The *Werke Nummer,* 2710, appeared in small white numerals but without the usual prefix, W Nr.

ing. The latter appear to have retained their overall dark green finish with pale blue undersurfaces, although possibly the ones used by JG 71 may have been repainted. In both cases, standard *Balkenkreuze* were applied to all six positions and a plain black *Hakenkreuz,* edged narrowly in white, was painted on each side of the rudder. In addition, the white band and top wing centre section markings were added along with the identifying black numbers.

It would perhaps be of some assistance at this juncture to examine briefly the *Luftwaffe's* training school system and its subdivisions. By 1939 the training scheme had achieved a degree of quality unexcelled by any other European air force. Flying training commenced at an *A/B-Schule* where the pupil received 100 to 150 hours flying instruction on elementary types such as the Bü 131 and Fw 44 etc. For first solo flights, red streamers were attached to the wingtips as a warning to all and sundry!

Students selected for bomber or reconnaissance training then graduated to a *C-Schule* where they received about 60 hours advanced training on twin-engined types such as the Fw 58, as well as obsolescent service aircraft and current operational types. Aircraft built specifically for training usually conformed to the then prevailing standard finish of overall 63, but types seconded from front-line duties usually retained the camouflage scheme appropriate to their former duties.

Graduates then passed to a *Blindflugschule* (blind flying school) for instrument training where some 50 to 60 hours were spent on such types as the Junkers W34. To signify their rôle, and as a warning to other aircraft, two narrow parallel bands of 04 were painted around the rear fuselage.

The final stage of training took place at a *Sonderschule* (specialist school) for bomber and reconnaissance personnel where they were brought together as a crew and spent some three months on final instruction. Fighter pilots were accommodated at a *Jagdfliegerschule* where they received the last of their training.

Photographed at a training school near Vienna in early 1940, this Fw 44 displays the four-letter factory code that replaced the unwieldly WL-prefixed series. The lighter area along the fuselage side resulted from the removal of the old code and markings. The overall dull silver scheme with black wheel hubs indicates that the aircraft had been in service for some years. It also illustrates the fact that some aircraft escaped retrospective painting with the color 63.

An example of prevailing inconsistencies in training aircraft markings during the 1939/40 period. This Fw 44 bore the revised style four-letter coding, RN + AB, and the *Hakenkreuz* had been correctly re-positioned. The *Balkenkreuz* however remained unaltered, retaining the earlier-style narrow white border.

The Bü 133 provided a natural progression from the Bü 131 primary trainer. Because of the aerobatic qualities of the Bü 133 it formed part of the equipment of the advanced flying schools, principally those devoted to fighter training. The aircraft illustrated wore a standard finish of overall 63 and its military markings were correct for the period. The stylish red and white trim was a Bücker trade mark and many Bü 133s retained all or part of it during the first year or so of the war. The words *Bücker Jungmeister* were painted in block letters across the fin and rudder and repeated in black script across the vee of the trim. Note that all the interplane struts were painted black.

Operationally obsolete fighter aircraft were utilised by the *Jagdfliegerschulen* for advanced training. Because these aircraft were still potentially useful for front-line duties in an emergency, they were usually camouflaged. This Ar 68E wore a non-standard pattern of 70 and 71 on its upper surfaces with the undersurfaces painted 65. The broad white band, believed to be one metre wide, and bearing the aircraft's individual school number, obscured the last letter of the black four-letter code. Thus only RB + B remained visible on the port side, while on the starboard side the letter R was obscured. The aircraft's number was repeated in black on a white panel on the upper surface of the top wing centre section.

Pre-war vintage operationally obsolescent aircraft were usually repainted in some form of camouflage. These Bf 109Cs from the fighter training school at Vienna-Schwechat however had their existing camouflage modified to meet the revised standards in force from late 1939 to mid-1940; namely the restriction of upper surface colors to a strict plan view. The military markings were of correct proportions and size and were also correctly located. The broad white band was restricted to the rear fuselage on monoplanes.

This Do 17E shows the revised colors used for second-line aircraft as introduced in March 1940. A special directive issued for aircraft of advanced training schools required yellow markings to be used as a warning to other aircraft. This Do 17 had a rudimentary application of yellow around the front of the engine cowlings and spinners. The close tone match between the 63 section of the upper surface camouflage and the yellow of the special markings, suggests that the latter were 04 rather than 27. The unit's badge, an owl on a roughly painted white shield, is very similar to the one used by *Flugzeugführerschule* (B) 16. The codes BA + BF, painted in black, had been thinly edged in white, a practice not uncommon amongst second-line aircraft. Both fins show some evidence of partial repainting with 61, the roughness of the application indicating that it was done by hand.

In March 1940, a further revision of markings for second-line aircraft took place. Blocks of four-letter registration groups were allocated to the various manufacturers and these were applied to all aircraft leaving the factory. On the fuselage sides they were paired, two on either side of the *Balkenkreuz,* while beneath the wings they were subdivided further by splitting each pair into individual letters either side of each *Balkenkreuz*. The registration markings were not applied to the wing upper surfaces. These markings were usually applied in black but Heinkels appear to have had a preference for white for most of their multi-eingined products.

Those aircraft already in service were again retrospectively modified. Each school was allocated an initial block of registration groups and these were applied to existing aircraft regardless of type. Thus at least briefly, miscellaneous groups of aircraft from the same school could be seen carrying registrations from the same block sequences; eg PF + LS, an Ar 66; PF + LW, a Go 145; and PF + OD, a Fw 58 — all of which belonged to *A/B* 4 at Prague-Letnany in early 1940. All three aircraft retained their narrow-edged *Balkenkreuz* whereas new arrivals were now sporting the wide-edged *Balkenkreuz* and the repositioned *Hakenkreuz* on the fin.

The retrospective modification of existing school aircraft caused the inevitable endless variations. In order to apply the new lettering to the underside of the wings, it was naturally necessary to paint out the old *Balkenkreuze* and reposition them further inboard on the undersurfaces. In some instances this was done correctly, but the old-style *Balkenkreuz* was re-applied instead of the new type with the wide white border. In other instances the modification order was misinterpreted and the new registration letters were applied to the upper wing surfaces as well. Arado Ar 66 PF + LS was an example of the latter.

Further examples of the variations possible were the

Two photographs of a Ju 86E of the same school. This aircraft had been repainted in 70/71/65 to which had been added yellow markings. The latter illustrate the standard established for this class of training aircraft, ie a half metre yellow band around the rear fuselage and a solid yellow panel beneath the wings from the *Balkenkreuze* outboard to the wing-tips. The black codes TO + CV on the fuselage were almost indistinguishable against the dark 70/71 background. (The actual camouflage pattern was just discernable on the original prints and provides yet another example of the low contrast values of the two greens when recorded on certain types of photographic emulsions.) The military markings were all of correct type, size and proportions. 198, the aircraft's *Werke Nummer,* appeared in small white numerals on the fin. Unlike the Do 17 BA + BF, the whole appearance of this Ju 86 with its correct and complete application of yellow markings and the neatly applied unit badge suggests that it was photographed at a later and more stable period of the unit's existence.

BA + ??, another Do 17E of the same unit, photographed at about the same time as the previous two shots were taken. It bore a scheme of 70/71/65 with yellow panels beneath the wings and, just faintly visible aft of the *Balkenkreuz,* a yellow fuselage band. The aircraft's neat overall condition seems to amplify the previous comments made about the Ju 86.

A detailed photograph of the owl emblem, this time adorning the nose of a Junkers W 34, BB + HC. The extent of application appears to have increased as the background shield had acquired an edging of a different color. The badge appeared on both sides of the fuselage.

An overall view of BB + HC showing the incorrectly proportioned upper wing surface *Balkenkreuze* which should have had narrow white borders. The yellow fuselage band is just visible immediately forward of the fin, half overlapping the letter C. This photograph helps to illustrate the fact that the yellow wing panels were confined to the undersurfaces. The overall finish appears to be a time-worn and weathered 63.

A 70/71/65-camouflaged Fw 58C school machine. It wore the usual yellow underwing panels and fuselage band, although the former did not extend right inboard to the *Balkenkreuze* on this particular aircraft. The code ?? + AW had been thinly outlined in white on the fuselage.

Unlike the multi-engined training aircraft, single-engined aircraft wore only a yellow fuselage band and, where fitted, a yellow spinner. This crashed Go 145 retained its old overall 63 finish and while the *Balkenkreuze* had been repositioned and updated, the *Hakenkreuze* still overlapped both fin and rudder.

An obsolete He 70F of a *C-Schule* shows the incorrectly proportioned *Balkenkreuz* beneath its wing. The aircraft appears to be finished in 70/71/65 but retains its Heinkel works name in white with stylish lightning flash emblem. On the original print it was just possible to see the yellow fuselage band which was overlapped by both the *Balkenkreuz* and one letter of its code. The front portion of the spinner was also yellow.

A good example of the many markings anomolies existing amongst aircraft of the training schools during the late-1939 to mid-1940 period. This Do 17E retained its original 61/62/63/65 scheme but the *Hakenkreuz* marking shows evidence of two stages of modification. Both fins had been repainted with 61 to obscure the old red band and white disc marking, but at a later date the *Hakenkreuze* had also been painted out and re-applied in the newer, correct location on the fins. The original underwing *Balkenkreuze* near the wingtips had also been painted out and each re-applied at the correct location half way between the centre line of the engine and the wingtip. The code letters SA + HN however, were too small in relation to the *Balkenkreuze* and the style of the letter N was incorrect. The combination of the fuselage code with the aircraft's school number was also unusual. The normal practice was to place the white number separately on the fuselage. Restricted space on the Do 17 seems to have influenced this particular application in an attempt to keep the number as central as possible. The use of white numbers and spinners for the aircraft of the *Sonderschulen* appears to have come into general use about Summer, 1940. A personal marking in the form of the name *Erika* in red appeared beneath the cockpit of this particular aircraft.

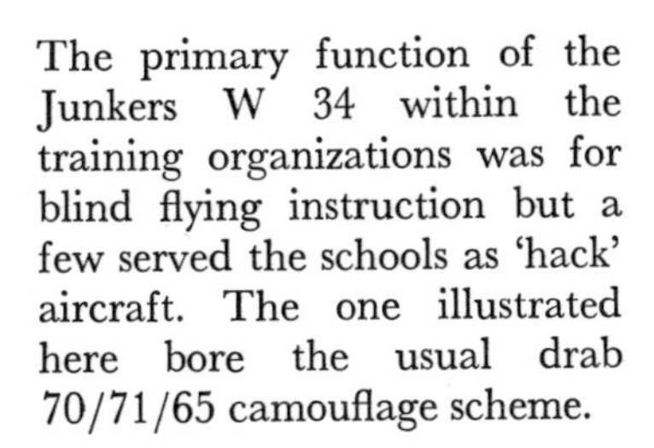

The primary function of the Junkers W 34 within the training organizations was for blind flying instruction but a few served the schools as 'hack' aircraft. The one illustrated here bore the usual drab 70/71/65 camouflage scheme.

Bü 133s of a particular *Jagdfliegerschule*. Since they were neither suited nor equipped for front-line duties in an emergency, they retained their very old overall 63 finish with the stylish red trim. As they belonged to a *Jagdfliegerschule* however, they bore a two-digit identity number. This was applied in combination with the new registration markings, the rearmost two letters being deleted in favor of the two numerals identifying the aircraft within the school. The rearmost two letters would have been deleted in any case had the Bü 133s been camouflaged and given the wide white rear fuselage band adopted by the other advanced fighter-training aircraft. The case in question is interesting inasmuch as the combination of markings was in use at the same time. If the numbers had been substituted *after* the four-letter registration, then the first two letters would have appeared forward of the fuselage *Balkenkreuz* on the port side and the last two letters of the registration group forward of the starboard *Balkenkreuz*. As it was, the letters LG appeared forward of the *Balkenkreuz* on both sides of the fuselage. The entire flight was marked in this way and ran from LG+01 to LG+10. These aircraft also retained the borderless plain black *Hakenkruez* in the old position overlapping both fin and rudder, together with the old-style *Balkenkreuz* with narrow white borders.

From late 1939, the *Jagdfliegerschulen* began to receive the operationally-obsolescent Bf 109D. These were marked with the usual school numerical identity aft of the fuselage *Balkenkreuz*. Since they had come from operational units however, they bore no factory codes. With the general changeover in markings to the new proportions and style *Balkenkreuz* and *Hakenkreuz*, the Bf 109Ds presented an interesting variation.

Both upper and lower wing *Balkenkreuze* had now been moved inboard to what was to remain, with minor variations, the standard position. The undersurface *Balkenkreuze* had the wide white border while the upper surface one retained the narrow border. Both retained the standard proportions but were larger overall. The fuselage *Balkenkreuz* was also of the new style with the wide white border.

The aircraft's identity number was applied in either solid black or white with a very thin black outline. The most significant change however, was the color of the rear fuselage band which was now yellow instead of the earlier white type. This was consistent with the revision to markings applicable to aircraft engaged on training duties. Spinners were also painted yellow at this time.

At least one *Jagdfliegerschule* operating the Bf 109D appears to have interpreted the instruction relating to the revised style of the *Hakenkreuz* in similar fashion to the staff of *A/B* 23 at Kaufbeuren. The latter replaced the existing type with a literally enormous type that spanned both fin and rudder and used most of the available space above the tailplane. The actual proportions however, were correct. Of the six photographs of Bf 109Ds from the *Jagdfliegerschule* mentioned above, all exhibited the incorrect size *Hakenkreuz* spanning both fin and rudder but again proportionally correct. This practice was by no means confined to the two schools mentioned.

Variety in markings was seemingly endless during this period of metamorphosis, but eventually it was to stabilize to a degree. Nevertheless it is one of the features of *Luftwaffe* camouflage and markings that the aircraft of the training establishments continued to provide a rich and colorful source of variety right up to the end of the war. For these reasons they will continue to be examined in detail in the following volumes.

A Ju 87B of a *Sonderschule* shows variation in the positioning of its white number. In this case the figure eight had been applied to both the engine cowling and the undercarriage spats. Like the Do 17 shown on the previous page, this Ju 87 had its own share of incorrect markings. The *Hakenkreuze* remained overlapping both fin and rudder and although the fuselage *Balkenkreuze* were of the correct proportions, those beneath the wings were of incorrect proportions but nevertheless correctly located. Light-colored patches beneath the extreme wingtips indicate where the original *Balkenkreuze* and aircraft letter had been overpainted. The four-letter black code was lost against the very dark 70/71 color scheme of the upper surfaces.

Transportgeschwader KG z b V 2 was formed in August 1939 and because it was held in reserve, did not see action in Poland. The five Ju 52s pictured here were all from 6 *Staffel,* the nearest two being coded G6 + JP and G6 + IP respectively. The aircraft identification letter on each appeared in yellow and is just visible, but the remainder of the codes, which were applied in black, are almost indistinguishable against the dark 70/71 background. Of interest to the discerning reader is the wingtip of what appears to be another Ju 52 in the immediate foreground. The contrast between the two upper surface camouflage colors is most pronounced, the combination being almost certainly 71/02. The equally marked contrast with the black of the *Balkenkreuz* should dispel any argument that the lighter colored area could have possibly been either 70 or 71. The *Staffel* emblem appeared on both sides of the noses of these Ju 52s.

CHAPTER 5

BLITZKRIEG!

At 1240 hours on 31st August 1939, War Directive No 1 was issued by the *Oberkommando der Wehrmacht,* and in the early hours of the next morning a flight of three Ju 87s from 3/St G 1 made the opening air attack of World War II. With war thus a violent reality, aircraft markings began a ferment of change, and whereas the August crisis of the previous year had placed the emphasis on defensive concealment, the swing to the offensive now produced a speedy reversal.

In the fast-moving war being prosecuted by Germany, rapid identification became essential with respect to both air-to-air and air-to-ground operations. The small wing *Balkenkreuz* had been designed principally so as not to compromise the aircraft's camouflage but unfortunately this proved a little *too* effective under battle conditions, and the vital seconds taken to distinguish friend from foe literally meant the difference between life and death for those involved.

The speed with which the Polish campaign progressed aggravated this situation as far as identification was concerned. By 3rd September the *Luftwaffe's* principle objectives had been achieved, and it was able to devote the major portion of its attention to the task of cooperating with the *Wehrmacht* in the destruction of the Polish ground forces. Even under the best conditions, aircraft operating against ground targets offer little

12
8
5
13
11

opportunity for rapid identification, and this problem lead to a hasty field modification of both upper and lower surface wing markings. The *Balkenkreuz* retained its proportions on the Bf 109s but it was enlarged and moved inboard to a point roughly half way between the wing root and the wingtip, at which point it was marked across the full chord of the wing including the flaps.

At first consideration the adoption of such a large marking for the wing upper surface also may seem unnecessary, but two points should clarify the matter. A steep turn at low level — the only practical way to change direction at speed and swing onto a ground target — inevitably exposes the aircraft virtually in plan view. Depending upon the relative position of the ground observer, it is possibly to see either the upper surfaces or the lower surfaces of the wings. The second point involves observation from higher altitude. Polish aircraft were certainly scarce at this stage of the fighting but they were by no means totally eliminated, despite assertions to the contrary in some books. Keepng in mind the very dark camouflage of the Bf 109s, so very simlar to that of Polish aircraft, and its effectiveness at low level, the danger of being bounced by other *Luftwaffe* aircraft was a very real problem. It is worth recalling that some 285 aircraft of all types were lost by the *Luftwaffe,* and another 279 badly damaged during the four weeks of this first campaign.

Other *Luftwaffe* aircraft engaged on low-level duties also adopted the revised wing markings. In the case of the Hs 126s the *Balkenkreuz* was only moved inboard a few feet to the point where the wing chord increased appreciably and allowed the marking to be applied as large as possible. Photographs of the Hs 123, the other principal aircraft type used in the ground-support rôle, are extremely rare for the Polish campaign, but from the few available for examination it would appear that they also used the enlarged form of *Balkenkreuz.* Unfortunately no similar corroborative evidence has so far come to light for the Bf 110s which were also used to some extent on ground strafing missions.

Significantly, the He 111s, Ju 86s and Ju 87s which all operated at much higher altitudes, and the Ju 52s which operated behind the German lines, retained the standard size wing markings.

Vague references have been made by some as to the use of temporary yellow coloring for tactical markings during this campaign. No evidence however has been found to substantiate this theory, and the field modifications mentioned would seem to refute the claim. A solid yellow nose cowling would certainly have been as effective as an enlarged *Balkenkreuz.*

Ambulance aircraft now found their pre-war mock exercises translated into stark reality in Poland. The ageing but seemingly ageless Ju 52 had been seconded to this task and its capacious interior made it ideal for the purpose. Most of the early Ju 52s used as aerial ambulances retained their existing 61/62/63/65 finish and markings, to which red crosses were then added. Initially these were displayed on a white disc forward of the fuselage *Balkenkreuz* and both above and below each wing. In addition, the aircraft's rôle was boldly identified by the legend *San Flz* (an abbreviation for *Sanitätsflugzeug*) followed by a number, also in white. Somewhat oddly, the aircraft's WL-coding was also applied in white on both the upper and lower surfaces of each wing. The red cross marking was used to

Opposite, upper. A Bf 109E of IV/JG 132 photographed in August 1939 displays an old boot emblem introduced by *Hauptmann* Johannes Janke and retained when the unit was later redesignated I/JG 77. This aircraft, white 13, carried the IV *Gruppe* white disc marking aft of the fuselage *Balkenkreuze.* The camouflage scheme was 70/71/65 with both the propeller spinner and the blades painted in 70. The pale-colored triangle just visible on the lower propeller blade was the VDM company badge.

Opposite, lower. A group of Bf 109Cs somewhere in Germany on the eve of war. In the background are aircraft of 2 *Staffel* of I/JGr 102 with their red identity numbers outlined thinly in white. The *Staffel* emblem, a caricature of a hunter with a blunderbuss, appeared on a large white disc beneath the cockpit. By contrast the aircraft in the foreground carried no *Gruppe* or *Staffel* emblem and thus their identity remains unknown. Their yellow identity numbers however, thinly outlined in black with lack of any symbol aft of the *Balkenkreuze* show that they belonged to 3 *Staffel* of I *Gruppe.* It is therefore possible that they belonged to I/JGr 102. Camouflage schemes and markings were consistent for all aircraft in the photograph with only two exceptions. The identity numbers on all aircraft were oversize, a carry-over from the earlier fighter markings system. Secondly, two of the aircraft carried rear fuselage band markings. The nearer of the two aircraft in question bore two full bands of yellow, thinly edged in black, while the other had a single band of red thinly outlined in white. This type of marking had appeared in limited use since 1935 and was presumably an aid to air-to-air identification. The aircraft carrying the number 13 helps to illustrate that the original establishment of 12 aircraft per *Staffel* had already been exceeded by late-1939. As the war progressed, numbers as high as 16 were to be seen, but they rarely exceeded this. Again the low contrast values between the 70 and 71 make it almost impossible to distinguish the actual camouflage pattern of the upper surfaces. The strong highlight reflections do, however, help to show that the paints used were flat and not matt.

When operations began in Poland the Ju 52s of KG z b V 1 were quickly supplemented by aircraft and personnel hastily gathered together from various operational training schools. This Ju 52 retained its old ventral turret but had its 65 undersurfaces repainted black when it was photographed in the autumn of 1939. This early use of black on its undersurfaces indicates that it saw service on the war front.

A Ju 87B being prepared for a bombing mission. Camouflage was the standard 70/71/65 but the division between the upper and lower surface colors was somewhat higher on these very early production B models. The spinner tip was painted in the yellow *Staffel* color.

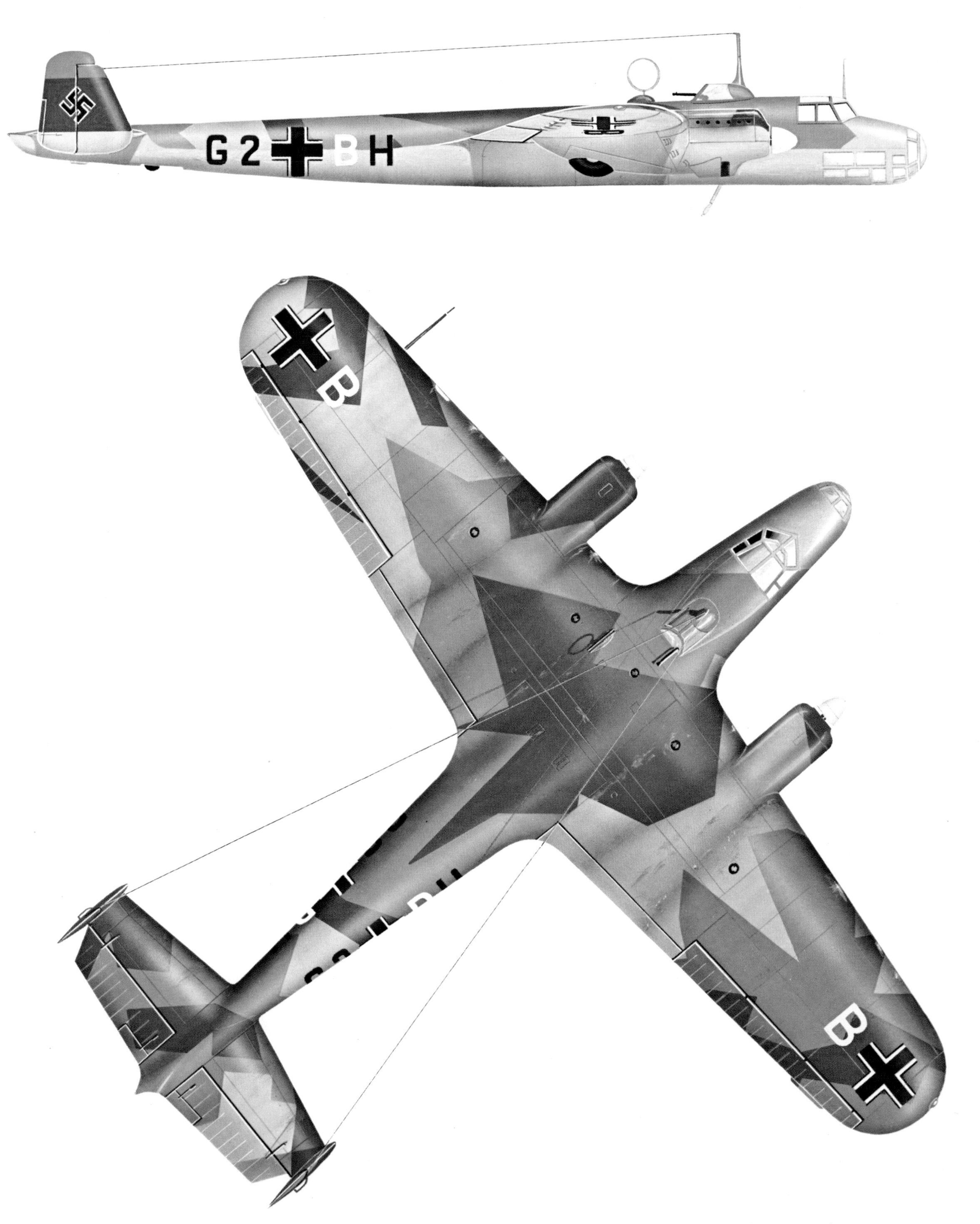

Still retaining its 61/62/63/65 scheme, this Do 17F-1 belonged to 1 (F)/Aufkl Gr 124 which saw action over Poland. Its markings complied with the new standards, the aircraft's letter being painted in the white *Staffel* color. The propeller spinners were plain white since the *Gruppe* color, normally displayed with the *Staffel* color, was also white. If comparison of the camouflage scheme is made with that shown on Page 68, it will be seen that the colors had been transposed.

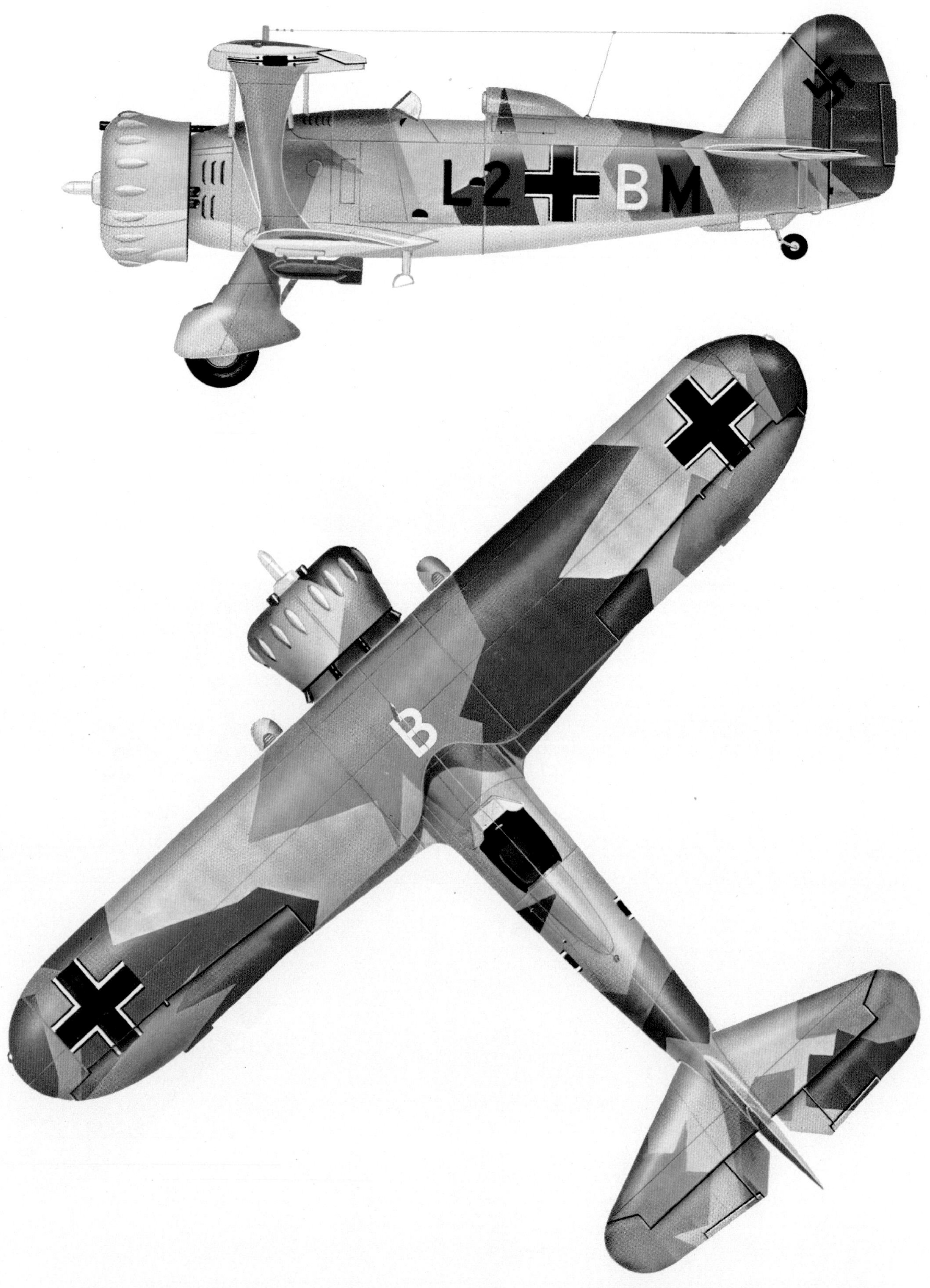

Another example of the practice of transposing upper surface camouflage colors. Compare this illustration with the one on Page 67. The only unit to operate the Hs 123 during the opening phase of the Polish campaign was II(S)/LG 2 which had 37 of its 40 aircraft serviceable at the time. The aircraft letter on this machine was applied in the 4 *Staffel* color of white. Note that the fin and rudder markings had been reduced to a basic black *Hakenkreuz* on each side.

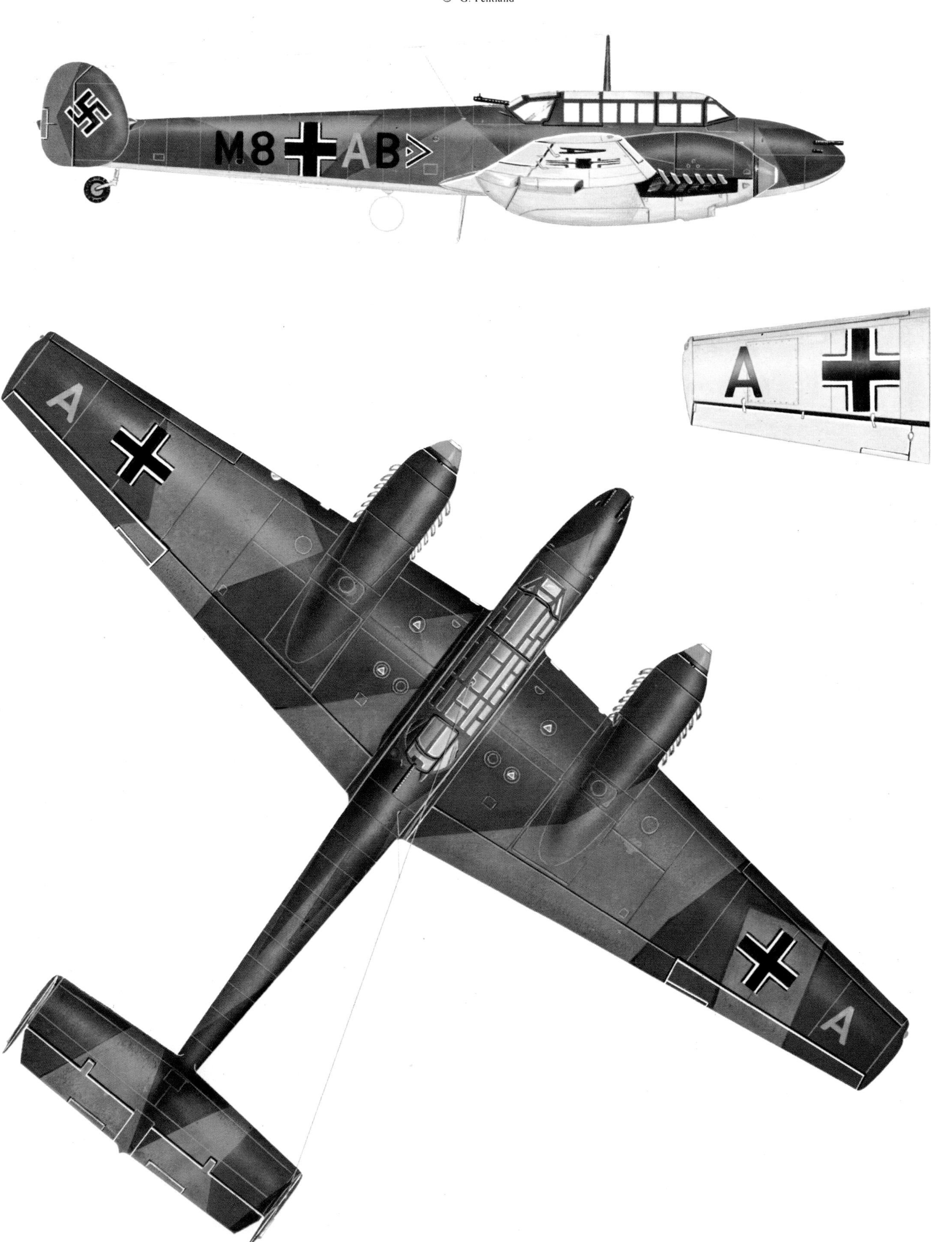

Only two front-line units used the Bf 110 over Poland in September 1939; I(Z)/LG 1 and I/ZG 76. This aircraft belonged to the latter unit which had 29 of its 31 aircraft operational for the opening phase of the hostilities. It was a staff aircraft as indicated by the letter B and the green coloring of the aircraft's letter and spinners. The white on the extreme tips of the spinners was the I *Gruppe* color. The double chevron marking denoted that the pilot was the *Gruppenkommandeur*. The color scheme was standard 70/71/65 and like many other aircraft of this period the *Hakenkreuz* overlapped both fin and rudder.

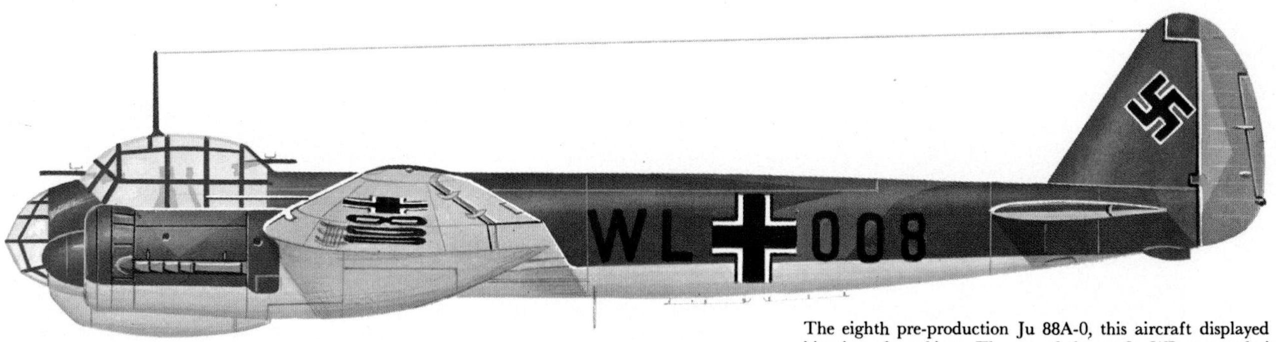

The eighth pre-production Ju 88A-0, this aircraft displayed an unusual com bination of markings. The use of the prefix WL was technically correct sinc the aircraft was not serving with a front-line unit. Instead of being used i conjunction wtih a civil registration however, it was combined with portio of the aircraft's *Werknummer*. The *Balkenkreuze* were of standard propor tions and correctly positioned while the *Hakenkreuz* retained only the thi white outline of this tense period. The rudder had been lightly oversprayе in 27, presumably for some reason associated with the aircraft's flight te programme. The general color scheme was standard 70/71/65.

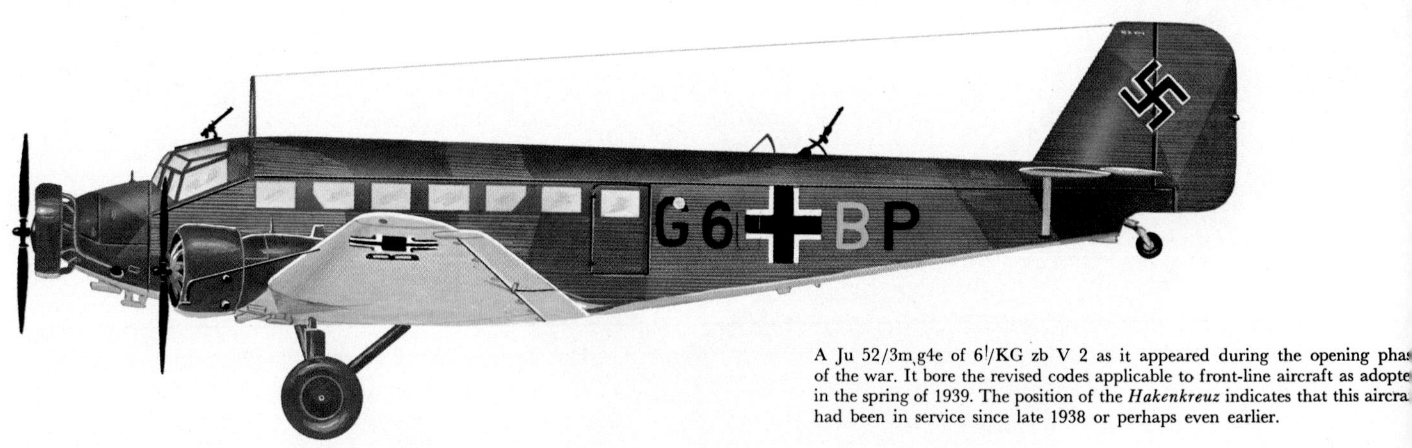

A Ju 52/3m g4e of 6./KG zb V 2 as it appeared during the opening phas of the war. It bore the revised codes applicable to front-line aircraft as adopte in the spring of 1939. The position of the *Hakenkreuz* indicates that this aircra had been in service since late 1938 or perhaps even earlier.

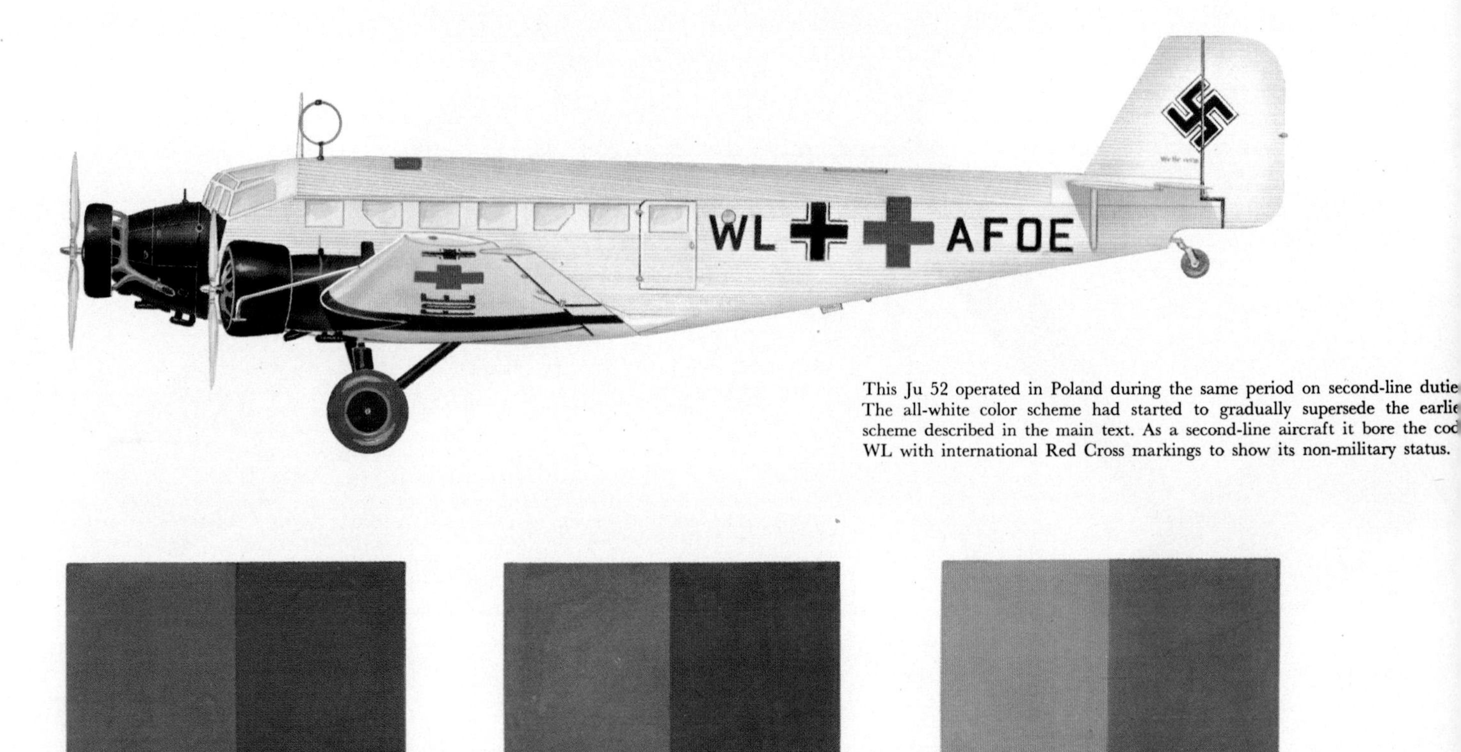

This Ju 52 operated in Poland during the same period on second-line dutie The all-white color scheme had started to gradually supersede the earlie scheme described in the main text. As a second-line aircraft it bore the cod WL with international Red Cross markings to show its non-military status.

Reproduced to show the extent of the contrast between respective combination these color samples, despite the limitations of modern high-speed printin processes, nevertheless provide a good idea of the difference in optical effe produced by certain upper surface camouflage colors. The left and cent samples illustrate the normal degree of variation between 70 and 71, the shad varying slightly according to the manufacturer and the particular paint batc The strong contrast between 71 and 02 will be obvious from the combinatic on the right. The colors 70 and 71 were an approximate match for the ligh and shadow areas respectively of European coniferous forests.

The following series of photographs shows the Ju 87Bs of I/St G 1 during the Polish campaign. Upper. This aircraft, A5 + GK of 2 *Staffel,* had its individual aircraft letter painted in red and thinly edged in what appears to be yellow rather than white. The use of white was more normal practice and would have been doubly significant since it was the *Gruppe* color. Lower. Groundcrew crank up the inertia starter of a Ju 87. The position of the *Gruppe* emblem, a cartoon-style diving raven, on the undercarriage spats was unusual. Note how the upper surface colors of 70/71 were brought well down the fuselage sides. The spinner tip carried the white *Staffel* color. In the background two other aircraft from 1 *Staffel* show just how well the white individual aircraft letters contrasted against the dark background. In the foreground is A5 + AH and behind A5 + CH, both of which carried their diving raven emblem in the more usual location beneath the front cockpit.

A

K

Upper. A photograph of interest to both the camouflage enthusiast and pure historian alike. A Do 17, probably an M model, is shown bearing the emblem of I/St G 1. The small rectangle in the lower left corner of the photograph was a manufacturer's plate. Made of aluminium, it was mainly black with the details pertinent to a particular aircraft stamped into the three bare metal horizontal strips.

Opposite, upper. An aircraft of the staff flight, in this case I *Gruppe,* bearing the code A5 + AB. There are several points of interest about this photograph; the actual code application is unusual in that the black letter A was marked lower than the accompanying number 5. Secondly the white letter A should have been painted in green. Instead, the aircraft's staff status appears to have been indicated by the white fuselage band which did not, however, continue underneath the fuselage. The machine's individual letter appeared inboard of the wing *Balkenkreuz,* but again in white instead of green, on the upper surfaces. The small rectangular white panel in the upper left hand angle of the *Balkenkreuz* was the access hatch to the emergency first aid kit, hence the small red cross marked on it.

Opposite, lower. A5 + KH of 1 *Staffel* shows the more normal location for the emblem just beneath the cockpit windscreen. The proportional relationship between the codes and the *Balkenkreuz* is clearly shown in this shot.

The cockpit of a Ju 87 showing the 7107-painted grey 41 interior and instrument panel. The black of the instruments stands out in distinct contrast with the grey background.

A Bf 110, M8 + DK of I/ZG 76, at dispersal in a pine forest somewhere in Poland. The style of the number 8, although unusual, was consistent for the aircraft of this *Gruppe*. The letter D was painted in the 2 *Staffel* color of red and was thinly outlined in white. The 70/71 upper surface colors were extended as low as possible down the fuselage sides for at this time the *Luftwaffe* was still primarily concerned with concealment on the ground.

A Bf 110 of 3 /ZG 1 which force-landed during the fighting in Poland. The *Staffel* emblem of a yellow outline of a lion's head with protruding red tongue can be seen on the nose section. Propeller spinners were yellow. In the background is a Do 17P, most likely of 2(F) /121, and a Fi 156. The effect of the light-colored rudders on the Do 17 was produced by the sunlight catching them as they were hard over to port when the photograph was taken.

A short-range tactical reconnaissance Fi 156 *Storch* photographed in the shelter of some trees on a forward strip shared by Aufkl Gr 10 in East Prussia. The effective visibility of the aircraft's individual letter, in this case O, is readily apparent.

Inspections and parades continued even in the face of war. A Do 17P of Aufkl Gr 10 *Tannenberg* photographed during the Polish campaign. The way in which the upper surface camouflage of 70 and 71 was swept down from the shoulder-mounted wings is clearly shown here.

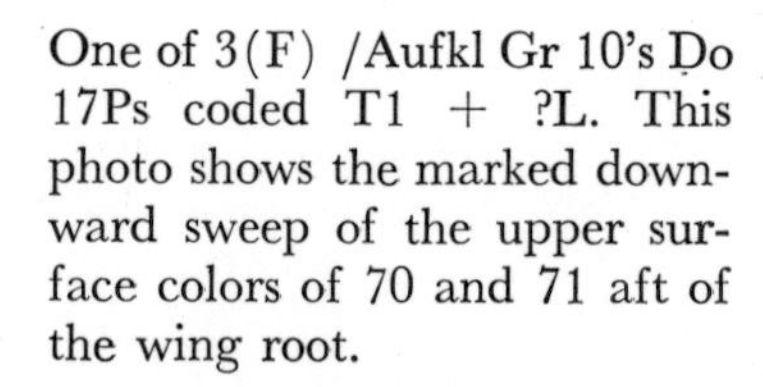

One of 3(F) /Aufkl Gr 10's Do 17Ps coded T1 + ?L. This photo shows the marked downward sweep of the upper surface colors of 70 and 71 aft of the wing root.

Left. A closer view of another aircraft showing the size of the code letter in relation to the *Balkenkreuz*. Lower. Aufkl Gr 10 also operated a short-range tactical reconnaissance *Gruppe* equipped with Hs 126s. They are seen here displaying the unit emblem of a black cross on a white shield. Note that the enlarged underwing *Balkenkreuze* had been moved to the inboard side of the aircraft's individual letter. This was a reversal of the normal marking requirement.

separate the WL prefix from the remainder of the registration and thus was applied at roughly mid-span.

This style of marking was in the process of being phased out in favor of an overall white scheme (as exemplified in the color illustration on Page 110), thus permitting a very positive identification of the aircraft and its purpose. The ravages of war however, required other Ju 52s to be seconded for ambulance duties and their rôle was identified by the simple addition of red crosses, albeit to the exclusion of some of the existing markings. Like most markings of expediency, the application was anything but consistent, and at least at this time, no definite standard appears to have been specified. This situation however, was to be remedied by mid-1940.

The *RLM* systematically analysed all aspects of the Polish operations and the field modification to the *Balkenkreuz* was now fully examined. As a result, an order was issued for the twofold modification of all underwing and fuselage *Balkenkreuze*. Firstly the proportions were altered to increase the visual identification range (see Diagram 19) and the actual positions of the wing *Balkenkreuze* were also changed as shown in Diagram 15.

Such at least was the intention. In practice this order was interpreted in a variety of interesting ways. Some Bf 109s could be seen with the lower wing surface national markings correctly repositioned but spanning the entire wing chord and using the old proportions. Others took the process a step further and also modified the fuselage crosses by increasing the overall size to virtually the full depth of the fuselage while still retaining the old proportions.

The application of huge wing *Balkenkreuze* was not limited to fighter aircraft alone, and examples were to be seen of both the old size being displayed along with the newly applied ones on wing upper surfaces. The He 111 illustrated on Page 128 shows one such example. At least one Do 18 took this type of combination a step further by displaying the lower wing surface style crosses full-span across its upper wing surfaces whilst still retaining the original type near the wingtips! No one could have possibly doubted its nationality.

On the Western Front, Winter of 1939/40 was cold, quiet, and generally boring for both sides. Activity was restricted to a few aerial reconnaissance missions and little else. In Germany the *Luftwaffe* continued the process of refitting and re-equipping.

Much has been said of the inconsistency of aircraft markings during the early period of the war, but inconsistencies were not restricted to markings alone. Both camouflage patterns and colors saw their share of variation although in most cases they were the exception rather than the rule. Some hybrid camouflage patterns were seen but these appear to have been instigated on a local basis. At this time the substitution of colors, and in particular one color, is both interesting and controversial.

It has never been resolved, and possibly never will be to the full satisfaction of everyone, as to why the color 02 should have been substituted for either of the existing camouflage colors of 70 or 71 in fighter camouflage schemes. What *is* known is that the practice certainly dates back as far as very late 1939. Suggestions for its adoption vary, the most common one being that it offered a more practical disruptive color scheme for fighter operations. This line of reasoning certainly has some merit, considering the type of operations undertaken by fighter aircraft as opposed to bomber types.

Another theory holds that groundcrews, when applying the aircraft's camouflage colors, utilised the 02-colored primer coat as one of the two upper surface camouflage colors. This theory however, has little practical merit when examined in depth. Aircraft under field repair certainly had new components, and any battle-damaged paintwork was soon repainted. It is pos-

JG 51's Bf 109Es were retained for home defence duties during the Polish campaign in case of sudden reprisals from other European countries. This aircraft wore the 7 *Staffel* bow-and-arrow emblem on its engine cowling.

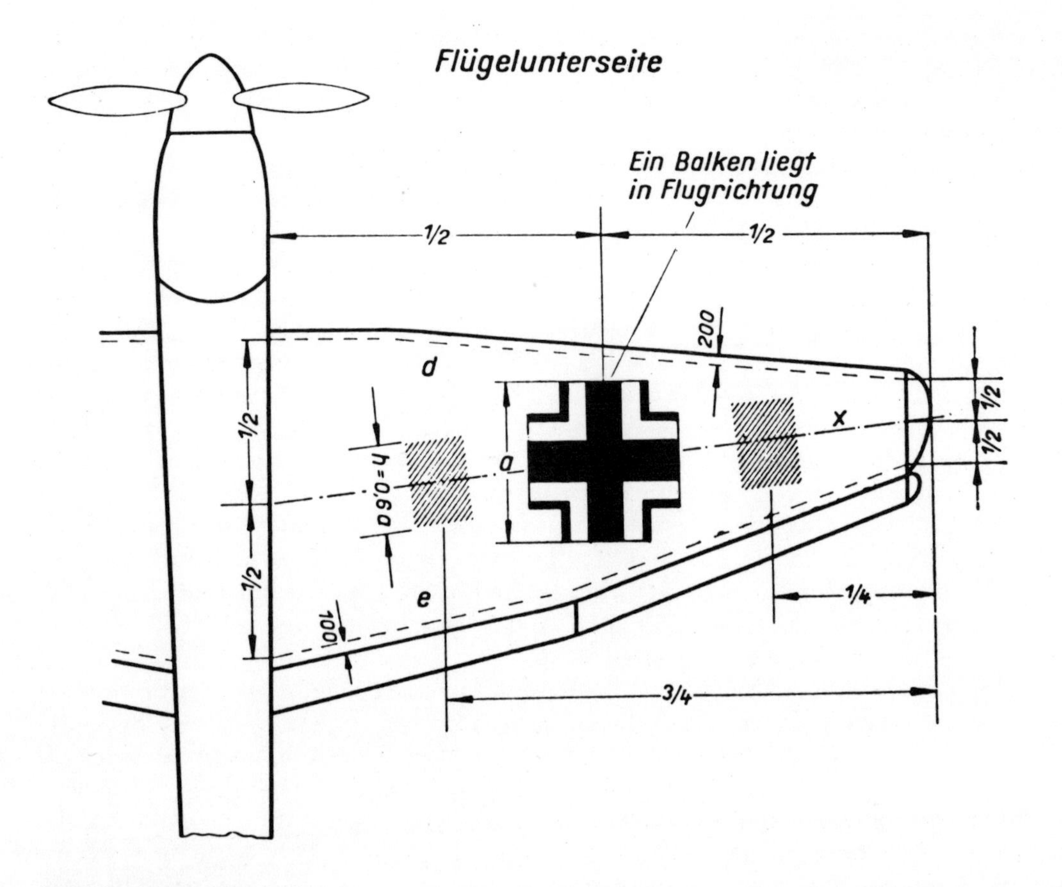

DIAGRAM 15

Location of military markings and codes on wing lower surfaces for single-engined front-line aircraft.

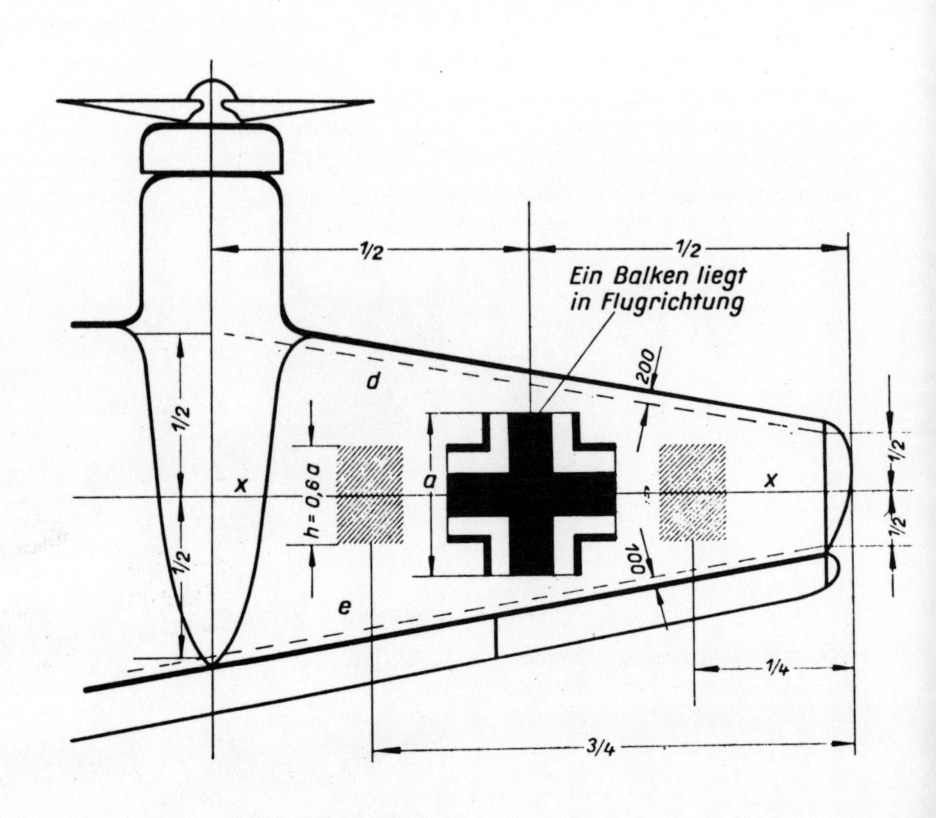

DIAGRAM 16

Location of military markings and codes on wing lower surfaces for twin-engined front-line aircraft.

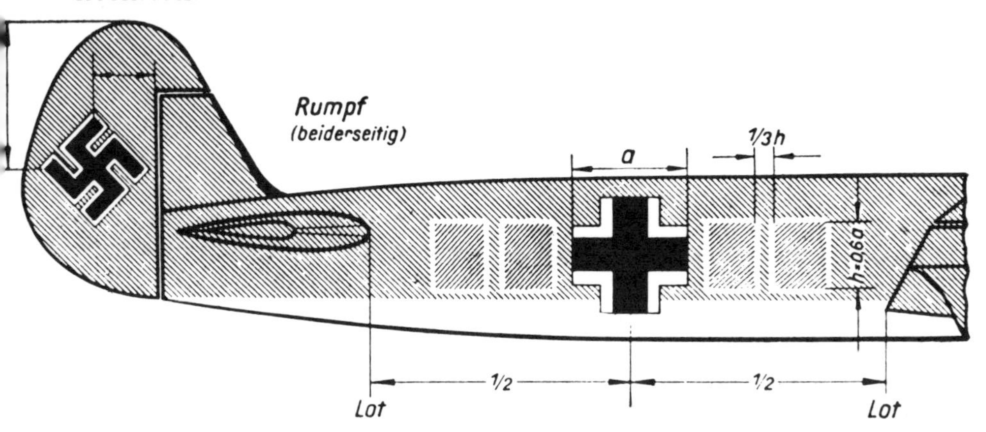

IAGRAM 17

ocation of military markings and codes on fuselage nd empennage of front-line aircraft. Positioning of the *akenkreuz* on the rudder was purely diagrammatic, e main implication being that the distance from the id-point of the marking to the tip of the fin/rudder as not to exceed two metres.

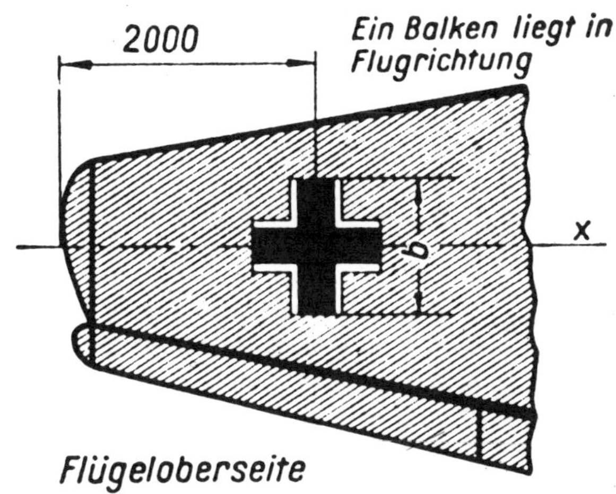

DIAGRAM 18

Location of the *Balkenkreuz* on wing upper surfaces of front-line aircraft.

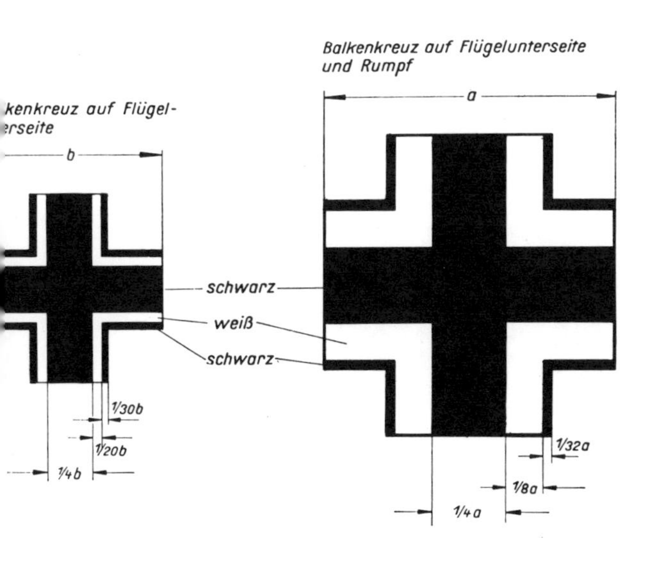

IAGRAM 19

oportions for wing upper and lower surface and fuselage *alkenkreuze.*

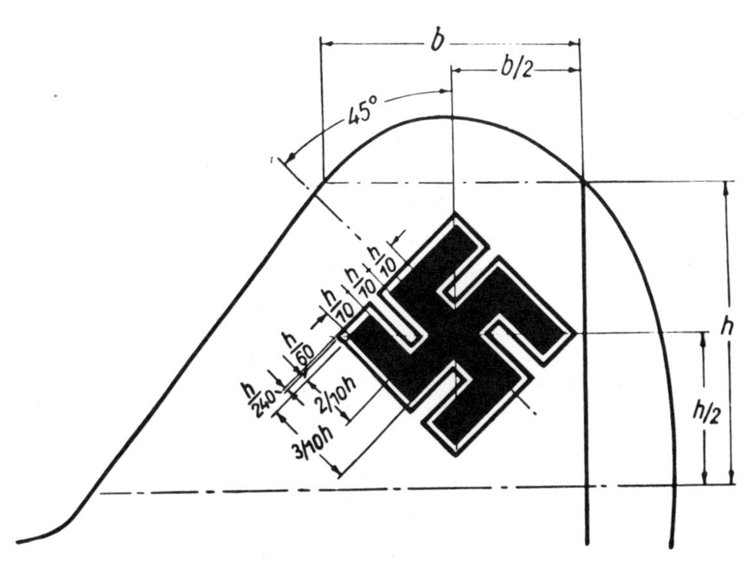

DIAGRAM 20

Accompanying the order to delete the red background and white disc to the *Hakenkreuz* was this official diagram specifying the proportions for the white border and black edging. While the position of the *Hakenkreuz* remained unaltered, the proportions were still related to the height of the original red background.

Another JG 51 Bf 109E, but this time exhibiting markings variations both standard and non-standard which reflected the operational experience gained in the Polish campaign by the *Luftwaffe*. The fuselage *Balkenkreuze* were of the revised type with increased width white surrounds. The upper wing surface *Balkenkreuze* were markedly non-standard. They had been moved inboard to the correct location but not only had the proportions for underwing *Balkenkreuze* been used, but they had also been applied across the full width of the wing chord. The rudder was not yellow, as would appear at first glance, for once again the angular displacement to the left had caught the strong sunlight which, in turn, produced a strong highlight sheen on the flat paint finish. The fuselage carried both the I *Gruppe* emblem just aft of the air scoop, and the skeletal hand 3 *Staffel* emblem beneath the cockpit.

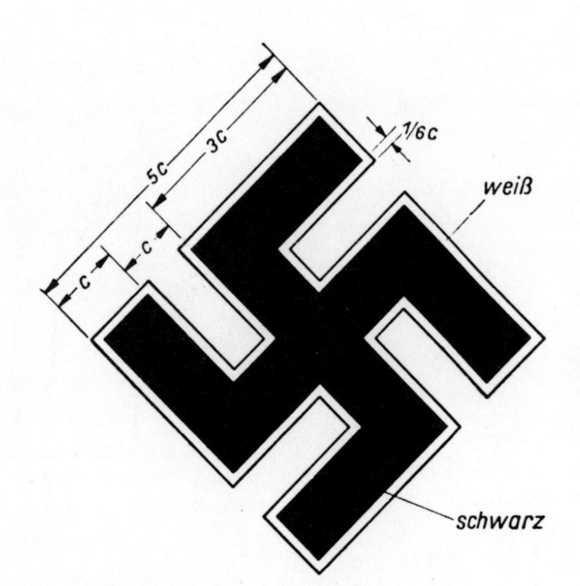

Abmessungen

Größe 5c	315	(400)	500	(630)	800	
c	63	80	100	126	160	
3c	189	240	300	378	480	
1/6c	10	13	17	21	27	

Auf Sichtschutzanstrich Farbton 70 ÷ 75 fällt schwarzes Innenteil des Hakenkreuzes fort.

DIAGRAM 21

The revised diagram issued when the *Hakenkreuz* was eventually ordered to be repositioned on the fin only. Exceptions were primarily seen in the products of the Arado and Fiesler firms as detailed in the text and photograph captions.

A Fi 156 communications aircraft used for second-line duties during the winter of 1939/40. Camouflage was the usual 70/71/65. The registration WL + IRWB appeared in black along the fuselage sides and beneath the wings. The positioning of the *Hakenkreuz* on the rudder was normal practice for Fi 156s.

sible that some of these new components were left in the basic primer coat of 02 but since aircraft were usually delivered to units already fully painted in the prevalent camouflage, it is a little difficult to see how the groundcrews could have become involved in painting the entire airframe. Aircraft were entirely repainted at certain intervals and to this end a small legend appeared on the port side of the fuselage which gave specific information. The following is an example from the period under discussion:

LACKIERUNG v. 1.7.40
(Lacquered on 1.7.40)
METALL: FL.7122.65/70/71
(Metal: Lacquer 7122.65/70/71)
HOLZ:FL=KETTE 30
(Wood: Lacquer group 30)
STOFF:FL=KETTE 20
(Fabric: Lacquer group 20)

Note: The literal translation *lacquer* has been used above whereas the more general term *paint* has been used in the main text.

The aircraft's log book stipulated a set period of time after which the machine's paintwork had to be reconditioned, for, as will be seen from photographs, the *Luftwaffe* was very particular about keeping its aircraft in good condition, even in operational areas. It was exceptional however, for an aircraft to be stripped back to bare metal and even then virtually impossible for only the top coats of camouflage color to be removed down to the 02-colored primer coat. Instead, repainting took the more practical form of an overspray of the existing finish which had first been cleaned down and reprimed where necessary.

One potential trap with *Luftwaffe* camouflage schemes is to confuse the color *pigment* with the actual paint *type*. The code 02 identified the pigmentation used in the primer coat but the actual paint finish was something entirely different. Paint types 7102 and 7122 were used fairly generally for both external primer and top coat finishes. With a primer coat of, say, 7122.02 and a top coat finish of 7122.70, it becomes apparent that to remove one from the other by some means other than laborious hand scraping would be far from practicable.

The earliest use of the color 02 as part of the camouflage scheme however, certainly appears to have originated as a local field experiment rather than as an official directive. Its use continued to grow with the

This group of four photographs of T6 + EP, a Ju 87B of II/St G 2, contains several points of interest. The yellow letter E followed by the letter P identified it as a 6 *Staffel* aircraft. Left and opposite. In accordance with the revised markings directive the fuselage and underwing *Balkenkreuze* had been modified and, in the latter instance, repositioned. The starboard wing however appears to have been minus its *Balkenkreuz,* as only a faint vestige of it is still visible. The old *Hakenkreuze* remained, overlapping both fin and rudder. The yellow spinner tip recorded the 6 *Staffel* ownership.

This shot shows the emblem painted just forward of the windscreen, a feature which can be seen a little more clearly in the lower photograph.

The actual emblem was Felix the Cat holding an umbrella; the mystery is that this device was first used by 8 /St G 51 and later by 5 /St G 1 but never, according to existing records, by 6 /St G 2.

Of additional interest is the makeshift attempt at field camouflage over the damaged aircraft while apparently being moved in stages to a place where better repair facilities were available.

Throughout the early months of 1940 the revised markings instructions were circulated to all units, albeit with some interesting variations. Factory-fresh aircraft appear to have conformed to the new standards with a fair degree of consistency. These two photographs show some Bf 110s awaiting delivery to front-line units. CF + NB shows the greatly improved visual effect of the revised style of fuselage *Balkenkreuz* and also the new location for the *Hakenkreuz* which was confined to the fin. Just to the left of the fin the first three white digits of the *Werke Nummer*, 301?, are visible. Behind was CF + NS.

A frontal view of CF + NS showing the factory code repeated beneath the wings. The F and the N, located between the engines and the fuselage, have almost been lost in the reflection from the snow on the wing undersurfaces. Early production Bf 110Cs tended to have their underwing *Balkenkreuze* applied a little too far inboard. According to the markings directives, they should have been located exactly halfway between the wingtips and the centre line of each engine nacelle.

A newly acquired Bf 109E standing on its nose helps to illustrate the revised size and positioning for the upper wing *Balkenkreuze* on single-engined fighter aircraft. Note also the revised upper surface camouflage limits which placed the emphasis on concealment in the air rather than on the ground. The *Luftwaffe* was very much a weapon of offence at this time.

A closer view of a newly issued aircraft, a Bf 109E, with its *Balkenkreuz* of revised proportions and 65-colored fuselage sides. The wing root had the walk area defined by a chordwise 28-colored line. Two separate 'no walk' signs can be seen, painted in what appears to be yellow 27. The white number two on the fuselage side is almost lost in the glare from the snow.

The Allied forces obtained their first close look at the Bf 109E when a machine from II/JG 54 crash-landed in France in April 1940. This general view of the aircraft shows that the old style camouflage and markings was still in use with front-line units.

A closer view of some of the damage and the II *Gruppe* emblem.

Gruppe and *Staffel* emblems ranged from the simple to the very elaborate. I/JG 77 displayed its emblem in generous proportions, no doubt to show the detailed workmanship involved. The aircraft's yellow number was thinly outlined in black and reveals its 3 *Staffel* status. This variation in numbering style was not uncommon during 1940. The fuel octane rating symbol, a yellow triangle narrowly edged in white with the figures 87 in black, contrasted strongly with the 70/71 fuselage sides.

Normally personal emblems, unlike *Gruppe* and *Staffel* emblems, were confined to the port side of the fuselage. The positioning of this cartoon-style illustration therefore suggests that it was a *Gruppe* or *Staffel* emblem but in such matters exceptions are always possible.

Left. He 111Ps under construction show the early 1940-style markings and manufacturer's codes. Both the Heinkel and Arado firms used white codes. Lower. Awaiting delivery. The striking contrast of the factory codes and military markings is quite apparent. Although white was use for the fuselage codes, black was always used underneath the wings for maximum contrast. One wonders if the second machine was a 'jinx' aircraft!

© G. Pentland

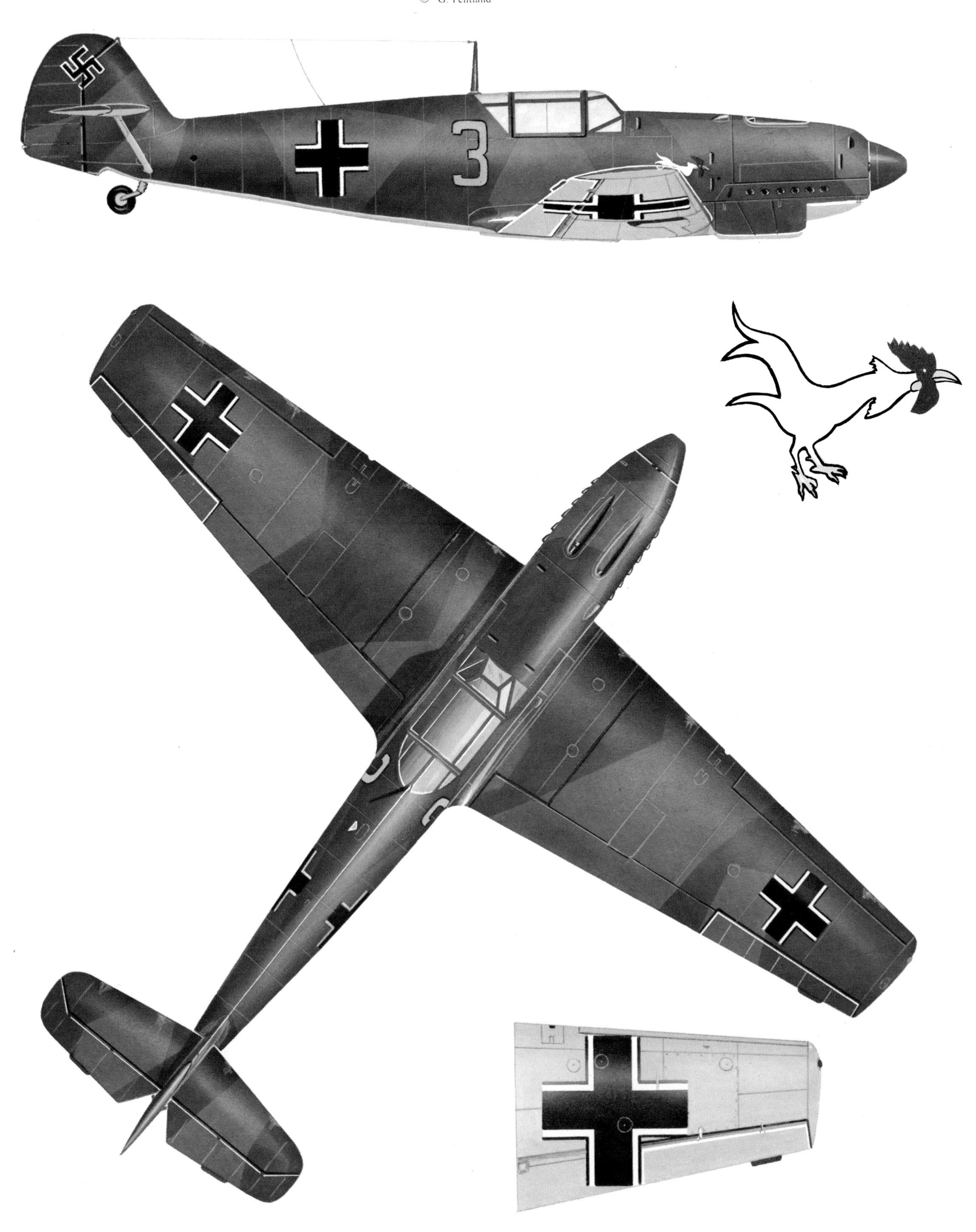

The Polish campaign produced a ferment of change in *Luftwaffe* military markings. The principal object of change was the *Balkenkreuz,* and an official directive was to set both new proportions and positions as explained in the main text. This Bf 109C of 9 /ZG 26 provides a good example of the initial changes instituted by some units in anticipation of the new directive. The wing undersurface *Balkenkreuze,* although moved inboard to the correct location, were grossly oversize. On the other hand the fuselage *Balkenkreuze* retained their old proportions. The *Staffel* emblem appeared on both sides of the fuselage.

Fighter aircraft were not the only ones to display incorrect markings. KG 26 had fought throughout the Polish campaign and carried out some of the few early daylight raids against English targets during the 'phoney war'. This particular aircraft from the Staff Flight was shot down near Edinburgh during the last week of October 1939. The full code group had been applied to both upper and lower wing surfaces, and on the fuselage also, in black. In view of the aircraft's special status its code letter should have appeared in blue.

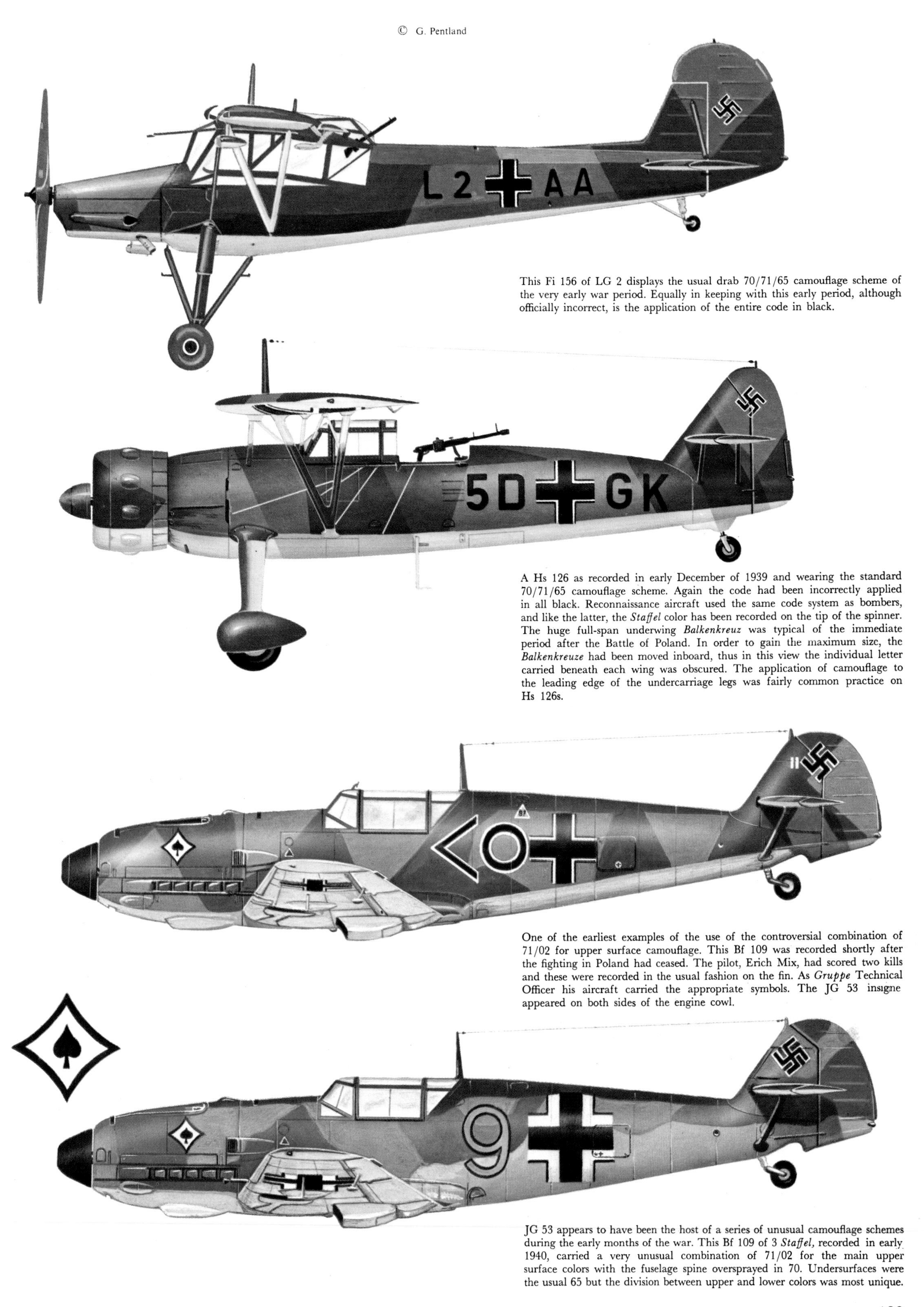

This Fi 156 of LG 2 displays the usual drab 70/71/65 camouflage scheme of the very early war period. Equally in keeping with this early period, although officially incorrect, is the application of the entire code in black.

A Hs 126 as recorded in early December of 1939 and wearing the standard 70/71/65 camouflage scheme. Again the code had been incorrectly applied in all black. Reconnaissance aircraft used the same code system as bombers, and like the latter, the *Staffel* color has been recorded on the tip of the spinner. The huge full-span underwing *Balkenkreuz* was typical of the immediate period after the Battle of Poland. In order to gain the maximum size, the *Balkenkreuze* had been moved inboard, thus in this view the individual letter carried beneath each wing was obscured. The application of camouflage to the leading edge of the undercarriage legs was fairly common practice on Hs 126s.

One of the earliest examples of the use of the controversial combination of 71/02 for upper surface camouflage. This Bf 109 was recorded shortly after the fighting in Poland had ceased. The pilot, Erich Mix, had scored two kills and these were recorded in the usual fashion on the fin. As *Gruppe* Technical Officer his aircraft carried the appropriate symbols. The JG 53 insigne appeared on both sides of the engine cowl.

JG 53 appears to have been the host of a series of unusual camouflage schemes during the early months of the war. This Bf 109 of 3 *Staffel,* recorded in early 1940, carried a very unusual combination of 71/02 for the main upper surface colors with the fuselage spine oversprayed in 70. Undersurfaces were the usual 65 but the division between upper and lower colors was most unique.

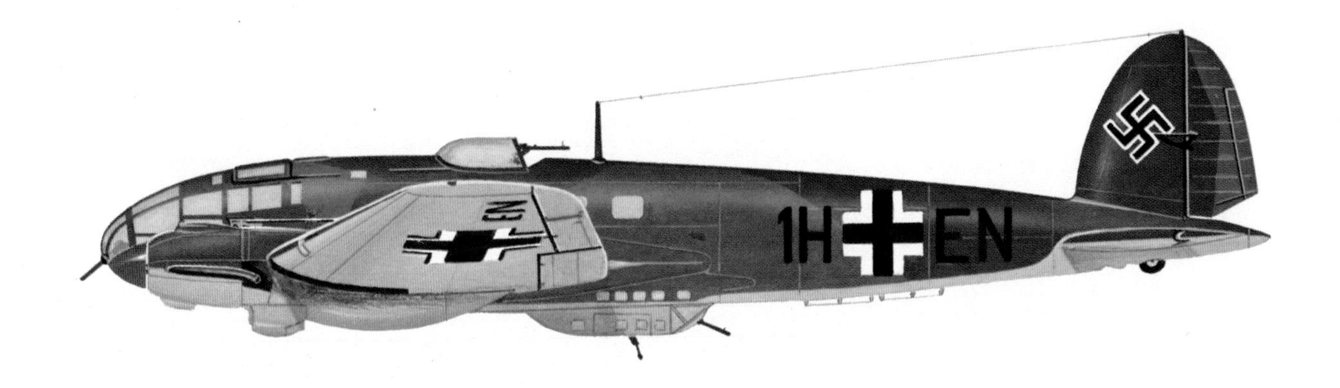

Recorded in February 1940, this He 111 of 5 /KG 26 makes an interesting comparision with the previous He 111 illustration. The markings had certainly been applied with a little more conformity but they still mirrored the minor confusions of early 1940. Although it was based on a normal mirror image reversal, the upper surface 70/71 color areas varied quite appreciably. The *Staffel* color on the spinners only was not uncommon.

The Do 215B-1 model was used by 3 Aufkl St /Ob d L during the Norwegian campaign. This aircraft had its 70/71 camouflage extended right down the sides of the engine cowlings. It is not known whether the damage to the starboard nacelle was caused by enemy action or an accident. This aircraft retained the older style fuselage *Balkenkreuze* with the narrow white edging.

A familiar photograph but worthy of inclusion since its camouflage and markings were anachronistic. The use of factory codes, CE + BN, dates the photograph at no earlier than March 1940 yet the aircraft retains the earlier camouflage with the 70/71 extending right down the fuselage sides. The fuselage *Balkenkreuze* and *Hakenkreuze* were of the correct style and in the prescribed location. The wing upper surface *Balkenkreuze* were, however, in the old outboard position.

The capture of the dual airfields at Aarlborg, in northern Jutland, were vital to the success of operation *Weserübung*, the assault on Denmark and Norway. Left. These Bf 110s of ZG 76 are shown in occupation, along with some Ju 88s and Ju 52s, at Aarlborg. Lower. The *Geschwader* code, 2N, is just visible on the aircraft to the left. Note the very light color of the interior of the cockpit framing where it had been folded down.

The Do 26 was doomed to a very limited production life. The V4 seen here was pressed into service along with the V2, V3, V5 and V6 as a *Sonderstaffel* attached to Kü Fl Gr 406 for long-range reconnaissance duties. Camouflage was presumably 72/73/65 and while the fuselage *Balkenkreuze* and *Hakenkreuze* were correct in all respects the wing upper surface *Balkenkreuze* were of incorrect proportions, having a wide white border. The black codes used were an odd mixture of military and manufacturer's applications. The first two, P5, were normal *Luftwaffe* style codes but the remaining two followed the normal factory allocation sequence, ie Do 26 V2 (DD), V3 (DE), V4 (DF), V5 (DG) and V6 (DH).

Old soldiers never die, or so the saying goes. These Bf 109Cs, despite their age, were pressed into service as part of the *Luftwaffe's* early night-fighter force. They wore the fully revised markings and camouflage scheme of the early-1940 period, their nocturnal rôle being denoted by the letter N. They belonged to 10 (N)/JG 77 and were based in Norway when this photograph was taken. In the background a Fw 200B, possibly of K Gr z b V 105, is just visible above N + 5.

passage of time and by late 1940 it was, from photographic evidence, apparently quite widespread for both single- and twin-engined day fighters.

The use of 02 as a camouflage coat was not restricted to simply replacing one of the existing colors, and on occasions it appeared as a random overspray in quite large patches. Some of the photographic examples examined of this type of application were recorded during the winter of 1939/40 and its effect for protective concealment was certainly appropriate. Possibly the actual paint type used was the temporary variety 7120.

Throughout the winter of 1939/40 the RAF kept up a series of incursions over German-occupied territory. The fact that a night defence force would inevitably be necessary had been foreseen by the *RLM,* but the priorities accorded this force were initially low. Thus a variety of older aircraft were seconded to these duties, among them the last front-line Ar 68s along with some Bf 109Bs, Cs and Ds. Camouflage schemes remained unaltered although some use was made of the temporary black finish 7120.22. Aircraft used for these duties were distinguishable by the special code allocation, a single letter N, which was used in conjunction with the normal fighter-style numerical sequence, the fuselage *Balkenkreuz* dividing the two.

In time the night-fighter force was to grow and to be subjected to a series of camouflage changes peculiar to its duties. For the period under discussion however, it underwent only one further change when the demarcation line between the upper and lower surface colors was raised, as noted later in this chapter.

The comparative inactivity on the Western Front at this time was soon to be shattered by the roaring might of a full *Blitzkrieg,* but there was first to be a Scandinavian prelude. The need to ensure the continued supply of vital iron ore from Sweden forced the German High Command to secure the long sea route which passed through Norwegian territorial waters. In addition, the occupation of Norway and Denmark was necessary to secure the German flank for the planned offensive against France and the Low Countries.

In the early hours of 9th April 1940 the long-awaited invasion commenced, Denmark falling on the first day, primarily to land and sea forces. The *Luftwaffe* was out in strength, but most of its aircraft overflew the Danish airspace en route to Norway. This was the first occasion on which mass air transport formations were used.

The Ju 52s were primarily drawn from units designated *Kampfgeschwader zur besonderen Verwendung (KG zbV),* (ie bomber squadron for special duties) and were thus classified as front-line aircraft. Painted in 70/71/65 camouflage, their code applications and colors were the same as those used by aircraft of the *Kampfgeschwader.*

The Norwegian campaign meanwhile continued to escalate and virtually the full range of available *Luftwaffe* aircraft saw action in those frozen northern skies. One of the newer types employed over Norway on a mixture of armed reconnaissance and supply missions was the Fw 200. These machines belonged to KG 40, a unit formed specifically for anti-shipping duties.

The elderly Do 17P model of this versatile bomber was still in use in 1940. This example, photographed during the spring thaw, displays standard camouflage and markings. The colored band around the spinner suggest that it had already been taken on charge by a front-line unit although no codes are visible except the letter M beneath the wingtip.

Unrivalled for tactical reconnaissance work, the Fi 156 served with the *Luftwaffe* throughout the war. These photographs show two aircraft which appear to have come from the same production batch. The upper one bore a white number 43 on its fin, the other 55. Both aircraft had incorrectly proportioned wing upper surface *Balkenkreuze* and an examination of 55 shows that all the *Balkenkreuze* had been modified by simply increasing the white border areas at the expense of the black edging. Significantly the *Hakenkreuze* on 55 were reduced in size to the normal dimensions. The camouflage scheme on both was 70/71/65. The light colored triangles on the wing roots of 55 were the yellow and white octane symbols for the wing tanks.

The Battle of France marked the end of the long waiting period since the beginning of the war. Combat engagements had hitherto been few with victories even fewer. Pictured here are three successful pilots of 1 Squadron RAF with some of their hard-won trophies. Holding a 71-colored Do 17 spinner is "Boy" Mould with "Bull" Halahan looking on. Displaying a yellow and black Aufkl Gr 122 crest is Leslie Clisby, first of the great Australian fighter aces of WW2. On 13th May 1940 he captured one of the biggest war trophies of all time and won instantaneous fame when after destroying three He 111s, he forced down another, landed his Hurricane behind it, then chased and arrested the crew at gunpoint.

May 1940 and the *Luftwaffe* prepares for its next major conflict. The camouflage scheme of this Ju 52 was 70/71 on the upper surfaces but the undersurfaces had been overpainted with 22-colored 7120, the temporary black finish. The *Balkenkreuze* were all of the revised type but use of the old style position for the *Hakenkreuze* still lingered on. Note the group of Do 17Zs in the background.

By May 1940 the Do 215B was in service with three *Staffeln* of Aufkl Gr/Ob d L, the major employer of this limited production version. The boldness of the markings of this machine, PK + EM, illustrates their effectiveness for rapid air-to-air identification.

Do 215B-4 PK + EH from the same production batch as the machine above displays the markings of the *Luftwaffe* of 1940—a victorious *Luftwaffe*. Note the practice of repeating the last two digits of the *Werke Nummer* on the rear fuselage and nose section just beneath the cockpit. This practice continued throughout the war years on several aircraft types but appears to have been somewhat random in application.

A Ju 87B of 9 St G 2 *Immelmann* has its wing gun reloaded during the French campaign. Generous quantities of mud on the wing testify to the tough conditions during the spring thaw. The yellow and black *Staffel* emblem is visible on the engine cowling.

Initial supplies of Fw 200s were painted in 70/71/65 but this was soon changed for the more appropriate 72/73/65 oversea scheme on later deliveries. The code style and application followed that of the normal land-based *Kampfgeschwader.*

As recorded in Chapter 4, the markings for second-line aircraft were revised in March 1940. After this date all aircraft leaving the manufacturers bore a four-letter code group on either side of the fuselage and beneath the wings. This replaced the old practice of simply marking the last two or sometimes three digits of the aircraft's production airframe number on the fuselage. On receipt at an operational unit, the manufacturer's codes were then painted out and the appropriate code group applied.

Coincidental with the introduction of this new system, the *Hakenkreuz* was repositioned on the fin such that it did not overlap the rudder. As always these revisions were intended to be applied retrospectively, but undoubtedly the pressures of the war kept groundcrews fully occupied with more important duties. Certainly the old location was to remain in use for some months to come. Home-based units not yet committed to the turmoil of daily operations carried out the orders with a reasonable degree of consistency.

By the time the main tide of battle broke loose on 10th May 1940, the *RLM* had already digested further lessons from its earlier campaign. The *Luftwaffe* was very much on the offensive and its old-style fighter camouflage, designed for defence on the ground, was revised accordingly. To provide more practical concealment for air-to-air operations, the dark greens 70 and 71 were restricted to virtually a strict plan view of the aircraft's upper surfaces, thus raising the color division line between upper and lower surfaces. Once again time and the pressures of war prevented this instruction being implemented simultaneously and the consequent variety was to be seen for some months yet to come.

New aircraft were in constant demand and the variety of camouflage schemes and styles was seemingly endless during the hectic days of "the push". Gradually, as replacement aircraft superseded older ones, the original bomber colors of 61/62/63/65 disappeared. Few if any examples of this style of camouflage were to be seen on front-line units by June 1940.

By the time the last battle-scarred vessel had withdrawn from Dunkirk bearing its precious human cargo, the *Luftwaffe* was glad of the opportunity to rest and refit its battle-weary units. Across the Channel the British hastened to fill the gaps in their defence network for the coming conflict that was now certain to pursue them from the war-torn fields of France.

The improved Do 17Z version formed a large portion of the *Luftwaffe's* equipment of early 1940. Opposite, upper. A machine from 7 /KG 3 photographed during the French campaign displays its *Staffel* emblem. All three *Staffeln* of III *Gruppe* used playing card emblems, the Ace of Hearts symbol being seen on the 9 *Staffel* machine beneath. 8 *Staffel* used the Ace of Spades and the *Stab* aircraft a combination of all three plus the Ace of Diamonds as in the upper photograph. On the latter aircraft the propeller hubs were painted green. Note that on all three aircraft the framing to the nose glazing had been painted to match the upper and lower surface coloring.

The ageing Hs 123 was in the vanguard of the fighting from Poland to France, giving valuable tactical support to the *Wehrmacht*. This close-up shot of a somewhat battle - weary machine shows the very high division line between the upper and lower surface colors, a feature common to many Hs 123s.

Communications and liaison duties were filled by a variety of aircraft types dependent upon the nature of the tasks undertaken. In the upper shot an Ar 169 photographed at Lille shows its overall 63 finish with a typical manufacturer's black four-letter code, VA + HP. At right is the same aircraft with small camouflage nets draped over the fuselage and wing *Balkenkreuze* to reduce the possibility of detection from the air.

A He 70F *Kurierstaffel* aircraft shows its 70/71/65 finish and prominent *Balkenkreuze* markings. On the fuselage, beneath the cockpit windscreen, the aircraft's *Werke Nummer,* 1795, appeared in white, a practice not uncommon with second-line aircraft.

A Bf 108 *Taifun* finished in overall 63 with four-letter code and revised style *Balkenkreuze*. The *Hakenkreuze* however, still overlapped both fin and rudder. In the background at the left can be seen portion of the 70/71-colored wing and tailplane upper surfaces of a He 111.

Aircraft serving with the *Luftdienst* were engaged on a multitude of second-line duties ranging from target towing to meteorological flights. Some 13 *Luftdienstkommandos* existed by the outbreak of war, each having around 120 aircraft on strength. The types of aircraft varied as widely as did the duties for which each was responsible. This Do 17Z carried the distinctive but small yellow and blue *Luftdienst* badge. In very small numerals beneath the badge was the marking 2/11. Several other *LD* badges have been noted bearing a similar style of marking. There is a possibility that this could have been a form of unit identification, eg in this case II *Gruppe* of *Luftdienstkommando* 11. The red and yellow spinners would then denote a 6 *Staffel* machine of II *Gruppe*.

Upper. The boldness of the revised style of *Balkenkreuze* is readily apparent from this photograph of a Junkers W 34. The color scheme was the usual 70/71/65 with black registration letters, RC + DF, denoting its second-line duties. Note that the F had been displaced slightly further forward due to the small window low down on the fuselage side. Lower. Another Junkers W 34, this time in use as a ferry aircraft. The prominence of the *Balkenkreuz* is quite clear. The low contrast of the 70/71 on the upper surface of the wing is also shown.

A *Luftdienst* aircraft, this time a Bü 131, finished in overall 63, a color that was to gradually disappear as stocks of this superseded paint diminished. The general finish of the aircraft with its black interplane struts, wheel hubs and undercarriage bracing struts was typical of the standard applicable to *A/B Schule* aircraft. It is possible that this particular machine had been taken over from a training school, the remnants of a badge being just visible in front of the centre section struts. Immediately aft of it was the Bücker company emblem and further aft and lower was the *Luftdienst* badge. Underneath that again were the letters WP in yellow. Most of the markings on W Nr 449 were correct for the period but the *Hakenkreuze,* although moved to the fin, were of incorrect and rather spidery proportions.

Obsolete types were pressed into service as transport aircraft and this He 111B is such an example. Its four-letter code, SE + GP, and revised style of fuselage *Balkenkreuze* help to date the photograph. The old 61/62/63/65 finish had been retained, a common practice where aircraft were received still wearing some form of camouflage. Note that yet again the *Hakenkreuze* remained overlapping both fin and rudder.

Two photographs of an obsolete He 111E-4 in use as a transport aircraft. Upper CH + NR shows its revised markings with the broad white surround to the black cross. Again the *Hakenkreuze* remained overlapping both fin and rudder. The color scheme was the original 61/62/63/65. Lower. This close-up shot reveals that one of the spinners had been replaced with one from another aircraft.

During the Norwegian campaign the Fw 200 was used for both anti-shipping and supply missions. By May the effective strength of 1/KG 40 was such that it was temporarily withdrawn from operations and re-equipped with the latest mark of aircraft, the Fw 200C-1. The fifth production C-1 is seen here wearing its black manufacturer's codes, BS + AJ. Although intended primarily for anti-shipping duties, some of the early machines were finished in 70/71/65 instead of 72/73/65. The apparent very marked contrast between the two upper surface colors on this aircraft is partly the result of highlights on the slightly glossy finish of one of the two greens used for the upper surfaces. Note carefully the area just aft of the cockpit as compared to the bottom line of the green camouflage aft of the wings.

At the front the battle continued in all its ferocity. Here a Bf 109E-3 displays the raised color division adopted earlier in the year. The *Balkenkreuze* had been modified but as so often occurred, the *Hakenkreuze* remained overlapping both fin and rudder. The unit identity remains unknown but the red 13 and lack of any symbol aft of the *Balkenkreuz* reveals that the aircraft belonged to 2 *Staffel* of I *Gruppe*. The pale-colored triangle on the propeller blade was the VDM badge.

Upper and lower. Two more shots of BS + AJ showing the other side of the segmented upper surface camouflage pattern. The degree of contrast between the two colors was still quite pronounced even without the strong sunlight to highlight one of the two greens.

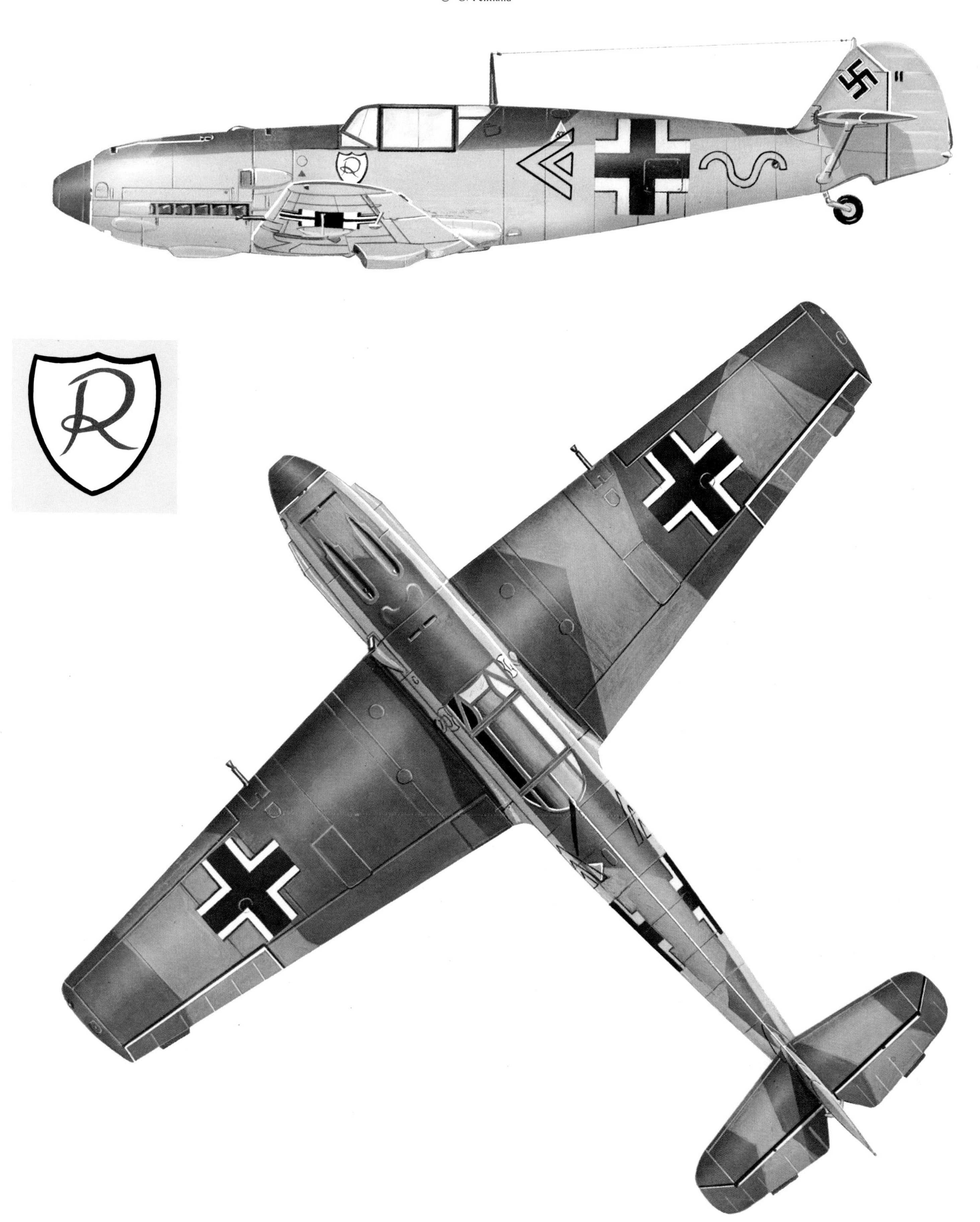

Erich Mix had risen to the position of III/JG 2 *Gruppenkommandeur* by May 1940. His previous mount, illustrated on Page 129, also carried the controversial 71/02/65 scheme but this one complied with the new directive regarding the raised demarcation line between upper and lower surface colors. Of interest is the use of outline-style *Stab* and *Gruppe* markings. In general this Bf 109E provides a good example of the main standards set for camouflage and markings by the closing stages of the French campaign. The script letter *R* device was the *Geschwader* emblem.

This Ju 88A-1 of 8 /KG 4 displays the drab 70/71/65 camouflage scheme typical of the period. Despite the revision of some of the markings, several inconsistencies were still to be seen. Upper and lower wing *Balkenkreuze* and the *Hakenkreuze* were all correctly located and marked, but the fuselage *Balkenkreuz* still retained its old proportions. The code also displayed the incorrect practice of merely outlining the aircraft's letter in white instead of using the red *Staffel* color. Instead, the latter was carried on the spinners, without the *Gruppe* color.

A Fw 200C-1 of I/KG 40 photographed in the spring of 1940. This particular aircraft was later shot down on the night of 19/20th July during a mining mission off Hartlepool. It was coded F8 + EH, the letter E being in the *Staffel* color of white. The highlight caused by the strong sunlight shows up the areas where the original factory code of BS + AG had been painted out. The much lower degree of contrast between the upper surface colors makes an interesting comparison with the earlier photographs.

The war at sea was by no means the prerogative of the Fw 200. The elderly He 59 continued to give service during the early months of 1940. This B-2 model, coded RC + BN, is thought to have been operated by KG z b V 108 *See* who utilised a variety of miscellaneous types. The camouflage appears to have been 72/73/65 but the very much darker area around the nose section is interesting.

The He 59 was also used for air-sea rescue duties, a task for which its low speed flying capabilities made it most suitable. This aircraft bears the white and blue emblem of 3 *Staffel* of the *Seenotgruppe*. The seagull was clutching a red life preserver.

The Ar 196 was a versatile reconnaissance aircraft and capable of giving a good account of itself in a fight. This A-2 model belonged to 5 /S A Gr 128 and like many other aircraft of this early period it did not use the *Staffel* color, in this case red, for the letter D. The marking of the *Hakenkreuze* on the rudder instead of the fin was characteristic of several Arado types as well as the Fi 156. The unit badge, positioned just above the wing root leading edge, appears to be very similar to that used by 5 Bord Fl St 196. The U-shaped markings on the fuselage were painted around the edges of the foot and hand holds. The 72/73/65 camouflage scheme was the same as for the He 59 in the background.

A Do 18G-1 of 2 /Kü Fl Gr 106 illustrates yet again the inconsistent nature of markings in the late 1939 to early 1940 period. The code M2 + ?K was incomplete, no aircraft letter having been applied, and the style of the letter M was unusual. The *Balkenkreuze* on the fuselage were also non-standard. The white edging had been increased but the black edging had been obscured in the process. The apparent colored engine cowl was, in fact, a close fitting canvas engine cover. The color scheme was 72/73/65.

KG 51 had completed the conversion of two of its *Gruppen* from He 111s to Ju 88A-1s by the time that the campaign in the west began. This aircraft 9K + EL carried the full quota of correct markings, the letter E being applied in the yellow 3 *Staffel* color. This color was repeated on the spinner tips. The camouflage scheme was 70/71/65.

A new and formidable breed of fighter aircraft was undergoing development trials in Germany at this time. Fw 190s were not to appear at the battlefront for another year but this test aircraft maintained the current camouflage standards—or almost. Originally registered D-OPZE, this was changed early in 1940 for the four-letter group FO + LY. The camouflage scheme was 70/71/65 but although the wing lower surface *Balkenkreuze* had been applied in the correct revised form and position, the proportions used were incorrect. The same problem applied to the fuselage *Balkenkreuze*. On the other hand the *Hakenkreuze* had been moved to the fin and correctly applied. The *Werke Nummer* appeared on the rudder simply as 01 in white.

New aircraft types were also the subject of further subterfuge. The German propaganda service was very thorough in its attempts to convince the British that a new type of single-engined fighter was not only in production but in front-line service. Their success was reflected by the combat reports that were later filed by some Allied pilots. The series of photographs shown here and on the opposite page were issued in the spring of 1940, and showed aircraft apparently in service. The camouflage was 70/71/65 and the *Balkenkreuze* were of correct proportions. Both upper and lower wing markings however, were a little too far inboard. The *Hakenkreuze* were also incorrectly applied to the rudder, a company characteristic with all the development models of this aircraft. The unit emblem was naturally a bogus one. The lighting symbol on the photograph opposite reflects a possible tongue-in-cheek attitude since it was very similar to Heinkel's pre-war trade mark. Most significant of all was the use of very high aircraft numbers. Few fighter units had in fact achieved an active establishment above 13 or 14 so early in the war. Note the contrast between the two wing colors on the photographic aircraft to the right of the upper photograph.

DIAGRAM 22
Do 17M AND P STENCIL MARKINGS

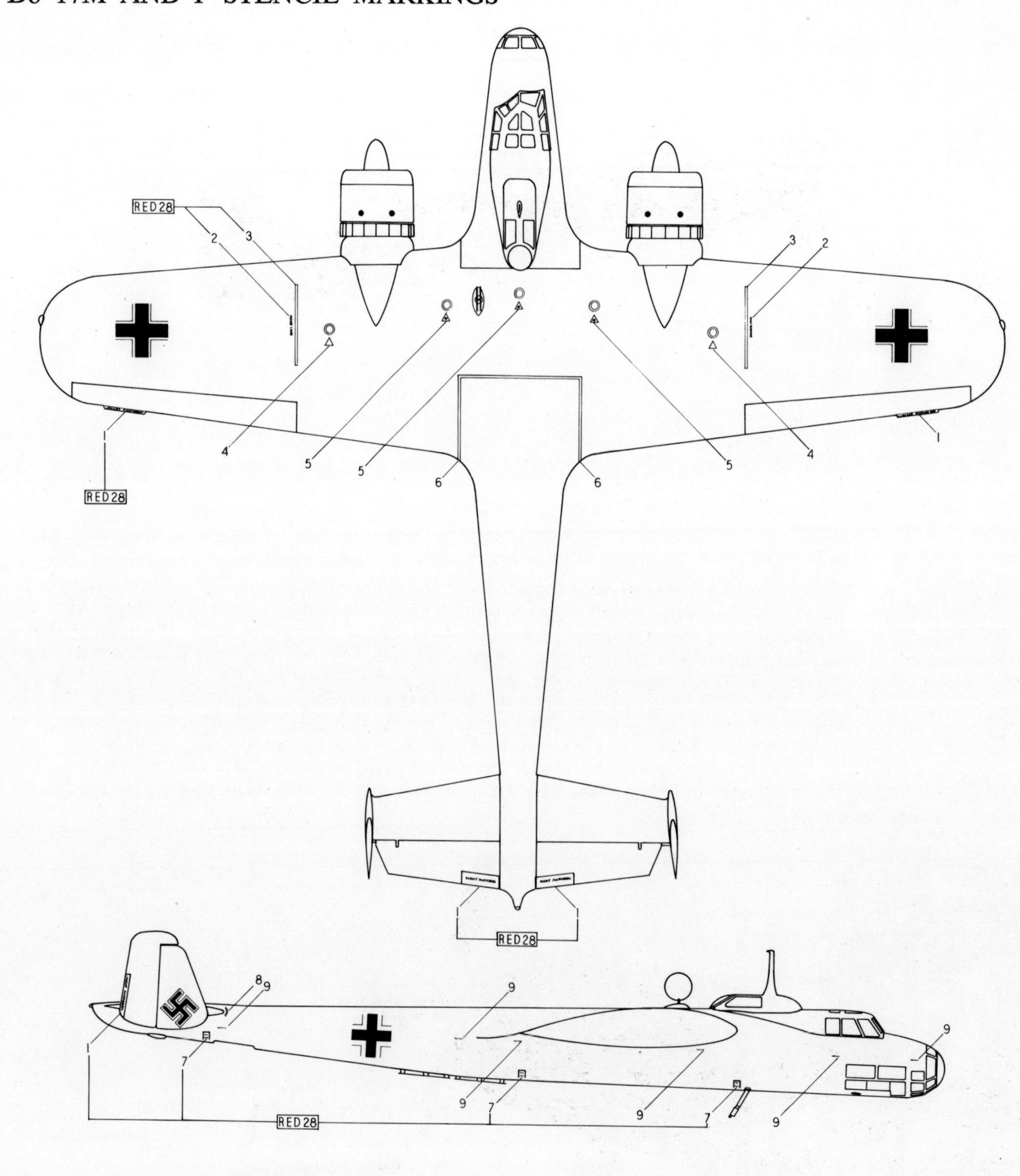

1 Nicht anfassen
2 Nicht betreten
3 40mm line
4 oil filler marking
5 fuel filler marking
6 white line
7 Hier unterbocken
8 taiplane adjustment scale
9 WE.(datum marking)
10 Verankerung

DIAGRAM 22
Do 17M AND P STENCIL MARKINGS

11 red-colored screw head
12 manufacturer's aluminium plate
13 Preßluft
14 red cross on white disc

all markings black except where noted.

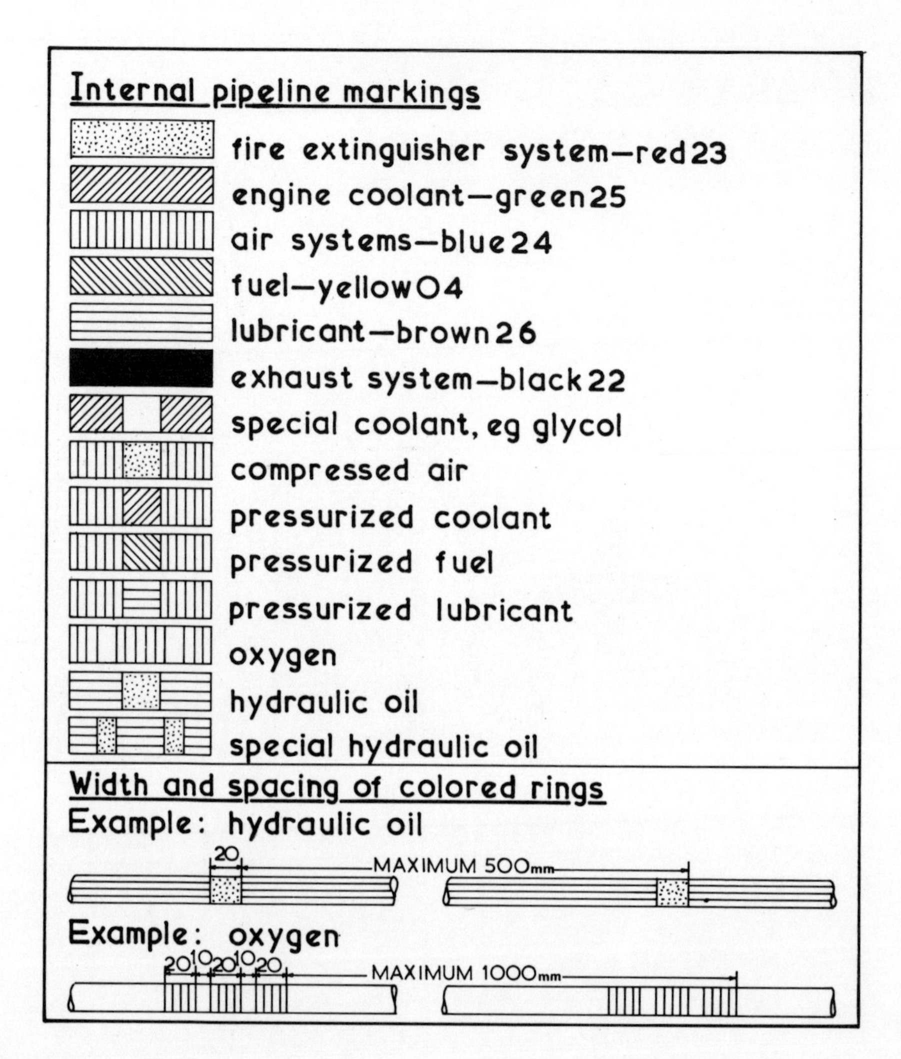

DIAGRAM 23

Color coding and dimensions for the identification markings on aircraft internal systems. Where a system had an external point, the relevant color code also appeared in order to identify it, (see Diagrams 22 and 25) either with a special marking or a thin band of color around the access hatch.

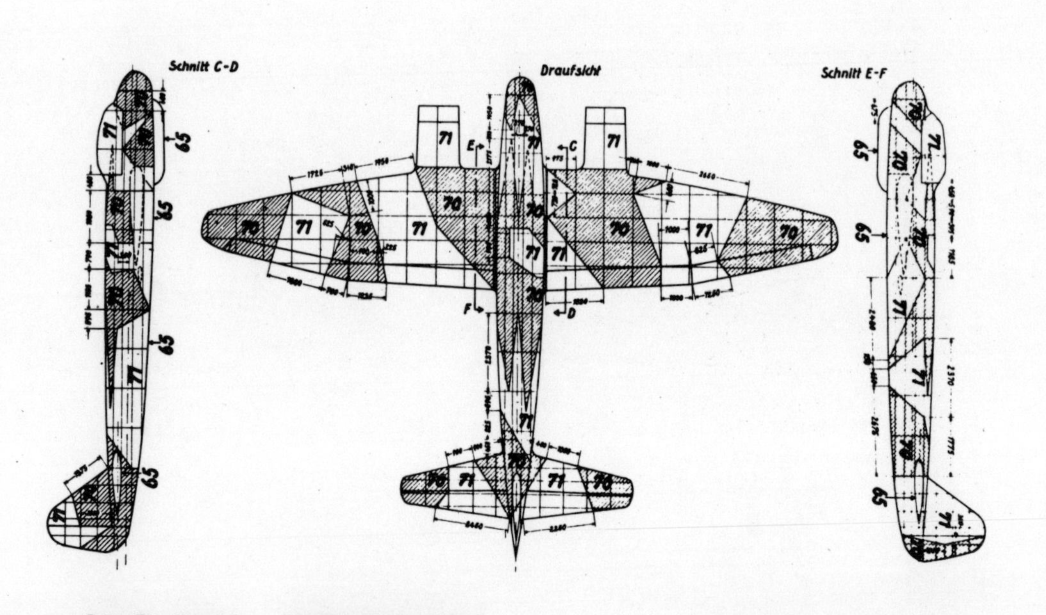

DIAGRAM 24

Left. Factory camouflage drawing for the Ju 88. The same diagram, redrawn to a larger scale and clearly showing the stipulated dimensions, appears on the opposite page.

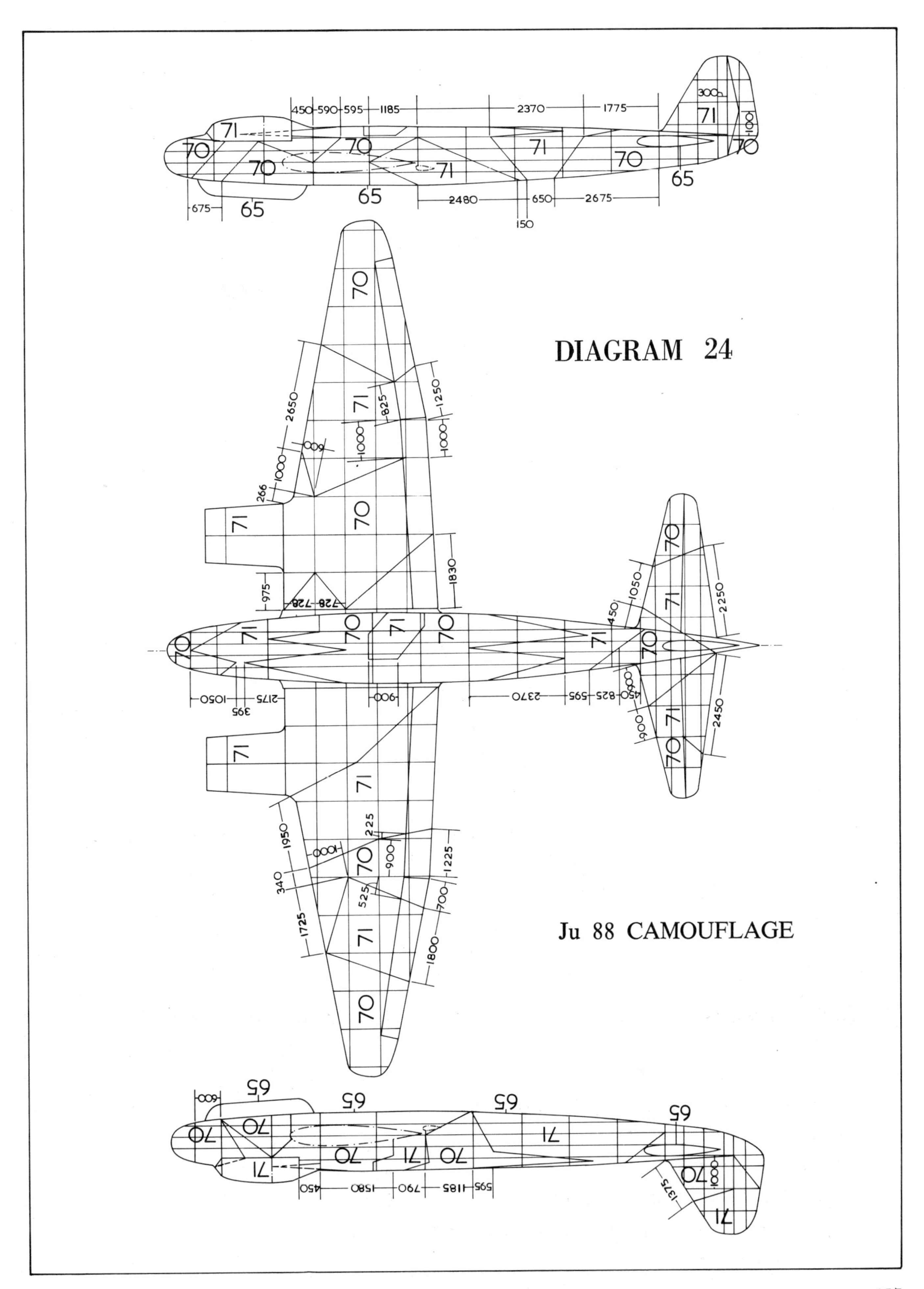
DIAGRAM 24
Ju 88 CAMOUFLAGE

DIAGRAM 25

Bf 109E STENCIL MARKINGS

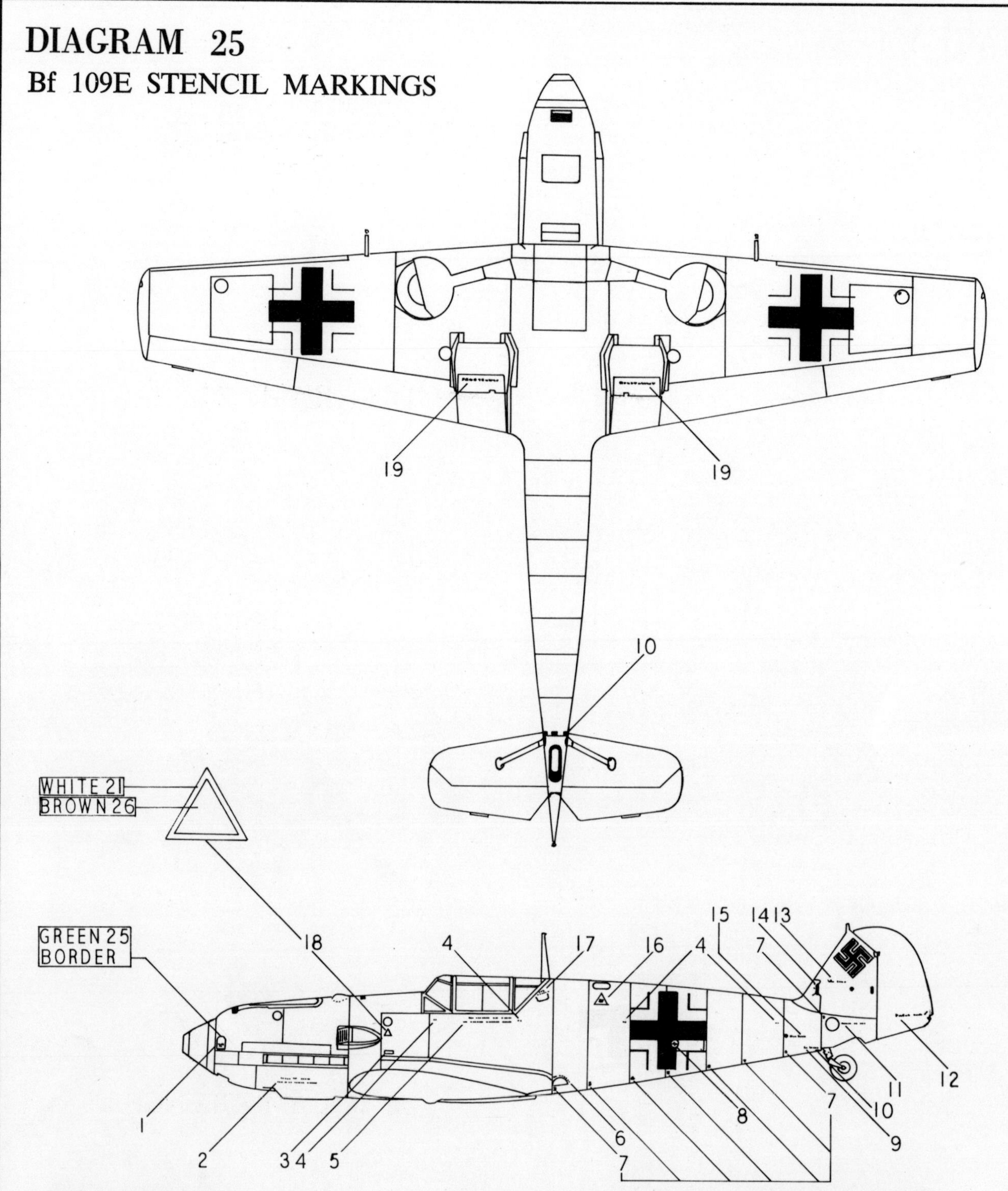

1 Glykol Wasser
50/50

2 Vorsicht beim Öffnen
Kühler ist im Hauberteil eingebaut

3 manufacturer's plate

4 WE.(datum marking)

5 Beim Schliessen der Kabine
auf Vorderteil Gummirahmen achten

6. Hier hintreten

7 frame numbers 2 to 9

8 red cross on white disc

9 Hier aufbocken

10 black rectangles

11 Reifendruck 4·5 atü

12 Nicht anfassen

13 W.Nr._ _ _ _

DIAGRAM 25

Bf 109E STENCIL MARKINGS

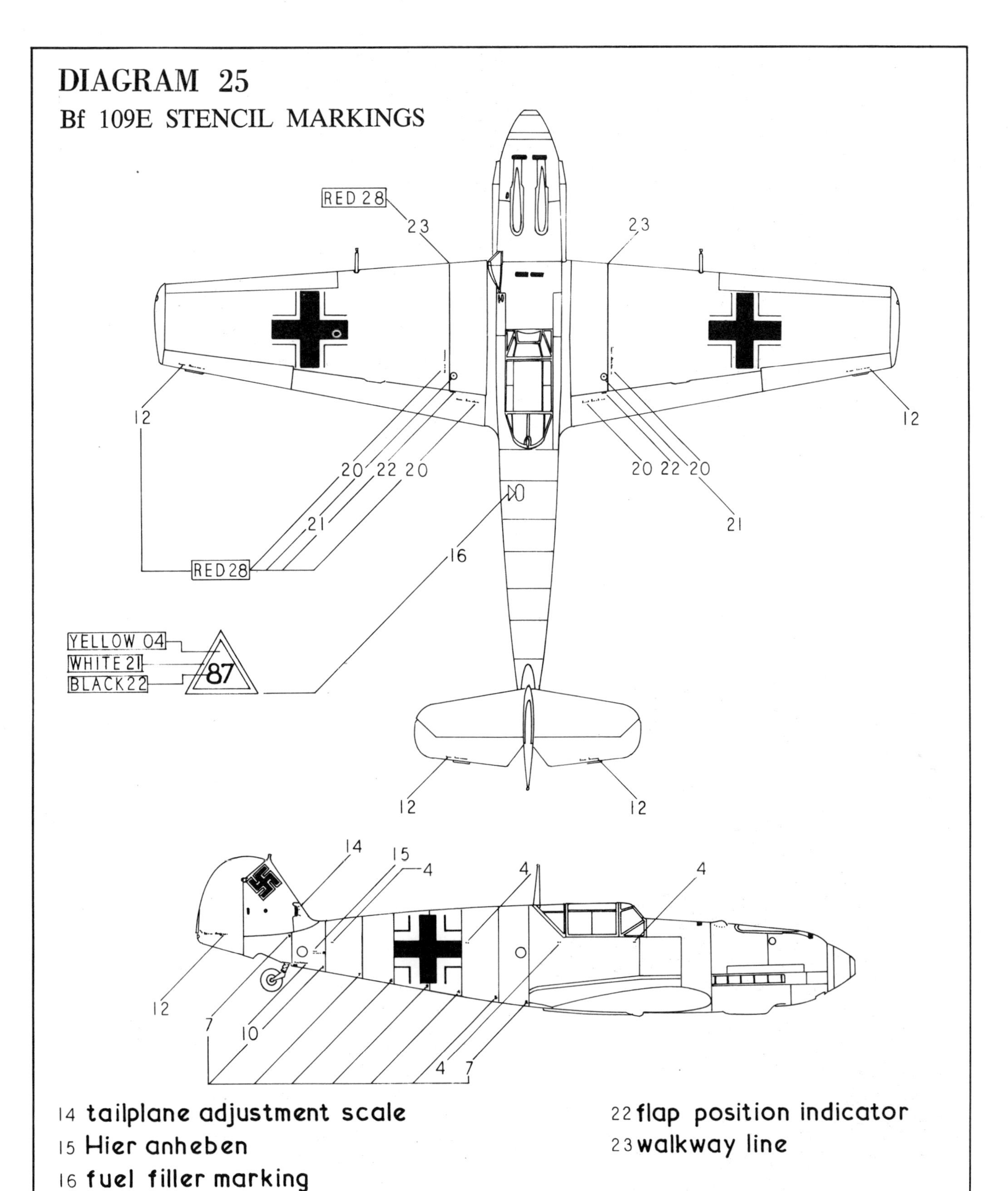

14 tailplane adjustment scale
15 Hier anheben
16 fuel filler marking
17 Hier eingreifen
18 oil filler marking
19 Frostschutzmittel
20 Nicht betreten
21 circle around u/c indicator peg
22 flap position indicator
23 walkway line

<u>all markings black except where noted.</u>

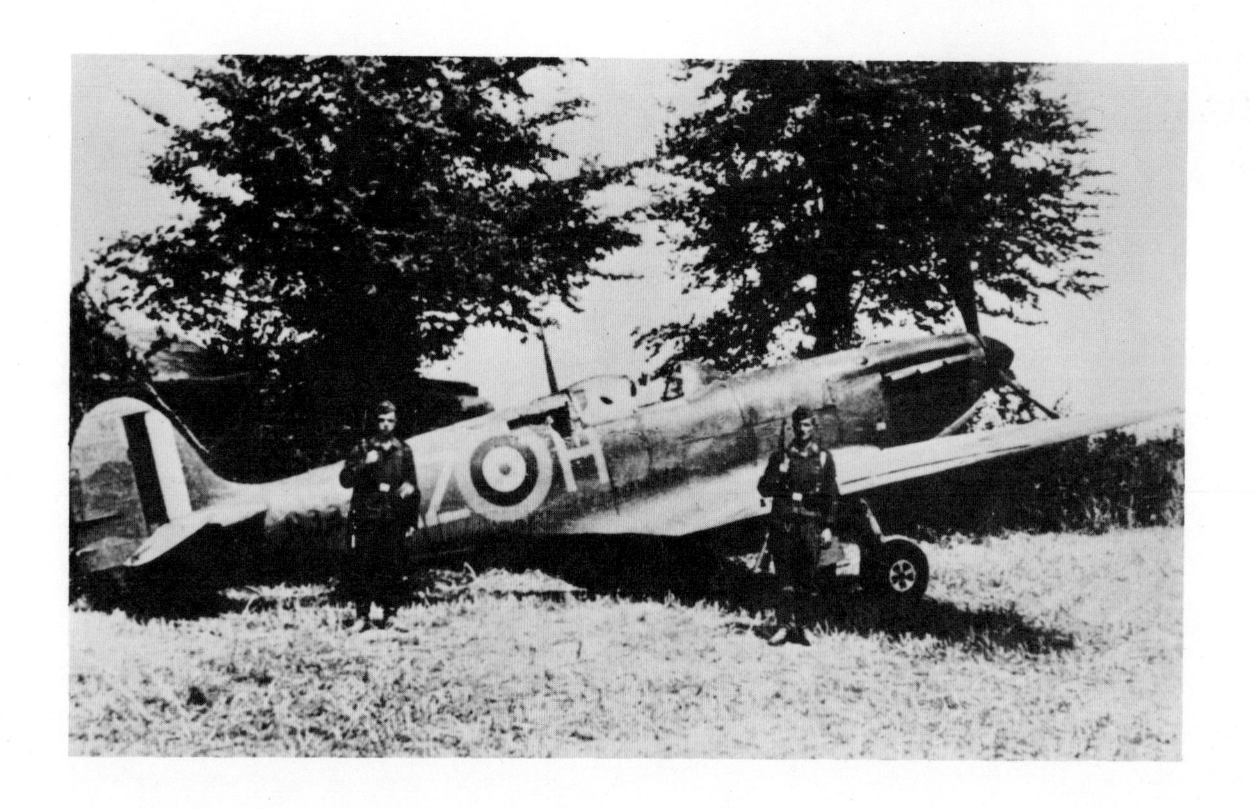

The early days of the fighting in the west were marked by losses on both sides. Pilots of the home-based RAF squadrons were desperate to come to grips with the *Luftwaffe*. Some did—and stayed for the duration of the war. These two photographs are somewhat anachronistic since N3277 did not actually force-land at Cherbourg until August 1940. They serve however to illustrate the point in question and show that personal emblems were not totally lacking in RAF squadrons. The words *Duty Dick* appeared on the starboard side along the engine panel line, just to the left of the guard's head. After the manner of the *Luftwaffe,* the personal emblem appeared only on the port side of the fuselage. Spitfire AZ-H, N3277 belonged to 234 Squadron. Shortly after these photographs were taken, a large solid black cross was painted on each side of the fuselage. The overall dimensions were the same as those of a normal *Balkenkreuz* of standard proportions.

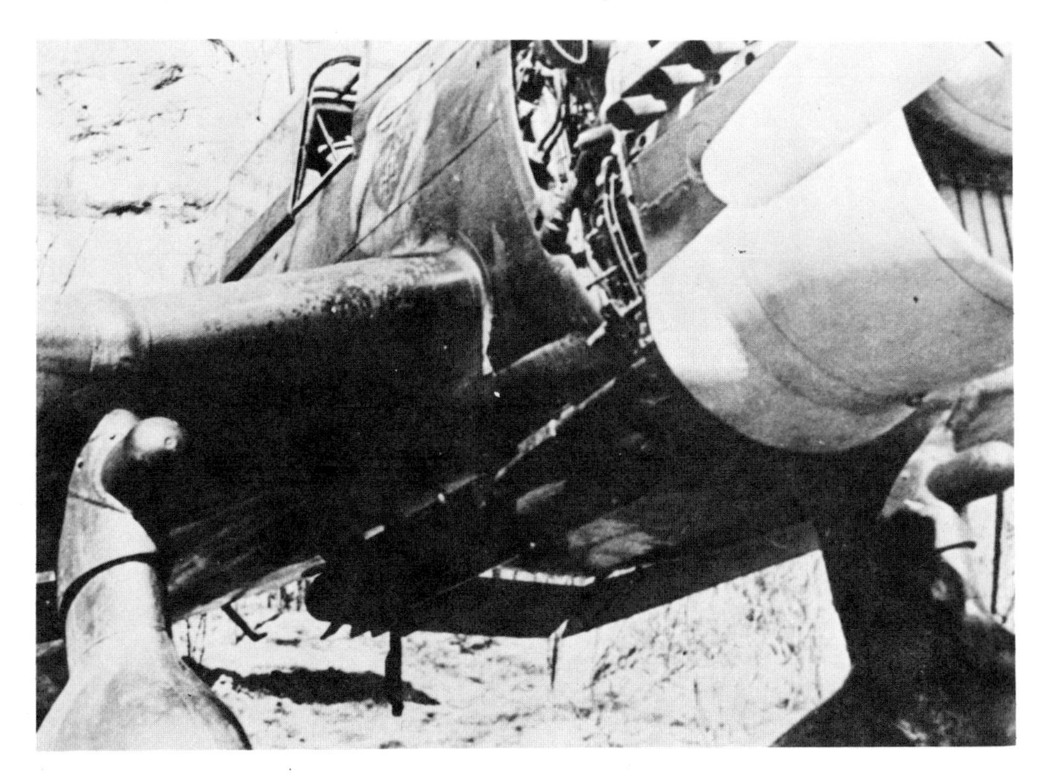

Upper. A Ju 87B-1 of 3/St G 2 shows the very large area taken up by its unit badge immediately above the wing root. Note the brightness of the 65-colored area beneath the radiator as compared to the exhaust-darkened surfaces further aft. Lower. A He 111 of an unknown unit displays the standard 70/71/65 finish for the period. The unit badge, a shield bearing an unknown device, appears to have yellow as its main background color. In some bomber units it was not uncommon for the main background color of the badge to be altered to match the particular *Staffel* color. In this instance this could have been the case since the spinners were also yellow. The code appears to have commenced with the number 2.

Reputedly the mount of the Staff Major of 2/St G 77, this Ju 87B-1 carried unusually elaborate markings for an aircraft of its class. The teeth were white with a bright red area between them, the eye being black and white.

Displaying the state of the art of camouflage and markings current by mid-1940, this Bf 110 of II/ZG 76 wore a 70/71/65 finish with fully revised and repositioned *Balkenkreuze* and *Hakenkreuze*. The aircraft's individual letter however, should have been applied in the 6 *Staffel* color of yellow, the full code being M8 + CP. Instead it appears to have been painted on in white, but as recorded earlier, this was not an uncommon fault of this period. What appears to have been a personal marking, a small light-colored heart-shaped device, can be seen just beneath the front cockpit quarterlight. Another small but significant marking was the line of three white kill bars on the fin.

Preparing for the next round. Such leisurely scenes as this were soon to disappear as the *Luftwaffe* commenced its all-out campaign against England.

It is perhaps appropriate to close this first book with a photograph of an immaculate Ju 52 finished in 70/71/65 and carrying the full standard markings for second-line aircraft as at mid-1940. The codes read VK + AZ. As will be seen in following volumes, the Ju 52 was to wear a great variety of interesting markings during the next four years of the war.

INDEX OF ILLUSTRATIONS